T4-ALG-591

THE GREEN REVOLUTION IN INDIA

THE GREEN REVOLUTION
IN INDIA

A Perspective

BANDHUDAS SEN

A HALSTED PRESS BOOK

JOHN WILEY & SONS
New York—Toronto

Published in the Western Hemisphere by Halsted Press,
a Division of John Wiley & Sons, Inc.,
New York

Library of Congress Cataloging in Publication Data

Sen, Bandhudas.
    The green revolution in India.

    "A Halsted Press book."

    1. Agriculture—India. 2. Agricultural innovations
—India. 3. Agriculture and state—India. I. Title.
S471.I3S412   338.1-0954   74-11066
ISBN   0-470-77590-4

PREFACE

The central purpose of this book is to present a balanced appraisal of the changes introduced in Indian agriculture by the high yielding crop varieties, popularly known as the green revolution. The real significance of these varieties has tended to be obscured amidst extravagant claims and scary forecasts of an impending catastrophe; what some have chosen to call the seeds of change, others have named the seeds of disaster. A balanced view of the nature of the changes brought in by these varieties, their achievements and failures, is needed before further opportunities for growth that are inherent in the situation can be identified and exploited. Much of this book is, therefore, necessarily concerned with sifting chaff from the grain and with the analysis of issues that appear to be the most relevant.

As this book grew from draft to draft, it became increasingly obvious that some of the conventional ideas about Indian agriculture and some of the conventional solutions to its problems needed reexamination. Owing to consideration of space, it has not been possible to do more than underscore these issues and raise questions about them. If these questions succeed in stimulating a thorough examination of all aspects of the issues raised, this book will have served a very useful purpose.

Some of the ideas expressed here were first put together tentatively in a paper, "Opportunities in the Green Revolution," published in the *Economic and Political Weekly*, in its Review of Agriculture, March 28, 1970. My grateful acknowledgement is due to S. V. Sethuraman and W. E. Hendrix for having read early versions of this book; and to Carl Malone whose comments on a summary of a few chapters circulated last summer were particularly stimulating. My sincere thanks are also due to Mrs. Grace Westerduin and to my wife for their useful editorial comments; and to Yog Raj Chhabra who typed the early drafts of this book.

However, the views expressed here are my own and should not be taken to represent those either of the reviewers or of the organization in which I work.

BANDHUDAS SEN

New Delhi
March 15, 1974

CONTENTS

Chapter One

INTRODUCTION

Nearly eight years have elapsed since the advent of the high yielding crop varieties on the Asian agricultural scene giving rise simultaneously to excited optimism and gloomy forebodings [1] Initially, there appeared to have been good reasons for both. The high yielding varieties are insensitive to differences in day-length (a property known as non-photo period sensitiveness) and early maturing; they give rise to short-stemmed, stiff-strawed plants that respond well to heavy doses of fertilizer in terms of grain. These attributes stand in sharp contrast to the characteristics of the local varieties. To illustrate, the local varieties of paddy are photo period sensitive, flowering and maturing when the day-length reaches a certain level after the retreat of monsoon; owing to this characteristic of delayed flowering and maturity and owing to the variability of monsoon, land could not be utilized to its maximum potential through multiple cropping or increased cropping intensity. Furthermore, the response of the local varieties to fertilizer is more in terms of vigorous vegetative growth rather than grain, leading to pre-harvest lodging and reduction in yield. In contrast, with appropriate levels of fertilization, yield of the new varieties can be increased by 50 to 200 per cent. These technical characteristics of the new varieties seemed to hold out a promise of a widespread use of these varieties and a substantial increase in the rate of growth of aggregate foodgrains production in underdeveloped

[1] For a chronicle of the circumstances leading to the technological break-through in respect of wheat and rice, see, Lester R. Brown, *Seeds of Change*, Overseas Development Council, Praeger Publishers, 1970. The beginning of the green revolution dates back to 1965-66 crop season when about 41,000 acres were planted under the high yielding varieties. See, Dana G. Dalrymple, *Imports and Plantings of High Yielding Varieties of Wheat and Rice in the Less Developed Nations*, U.S.D.A., February, 1972.

countries.[2] Self-sufficiency in respect of food, and what is more, export of foodgrains to earn foreign exchange had become all of a sudden real possibilities for countries that for long had been dependent on food aid and food imports on concessional terms.

These expectations were strongly reinforced by dramatic increases in acreage under the high yielding varieties and in foodgrains output in Asian countries. Area planted under the new varieties rose from a meagre 41 thousand acres in 1965-66 to about 50 million acres in 1970-71;[3] simultaneously, fertilizer consumption averaged an increase of 14 per cent per annum.[4] Growth of output in certain countries was equally impressive. Thus, wheat production in Pakistan rose by about 60 per cent between 1967 and 1969; during the same period, rice output in Sri Lanka increased by about 34 per cent; the Philippines exported some rice for the first time after a century of dependence on rice imports; and food self-sufficiency for India appeared to be right round the corner.

Initially, therefore, the new varieties were seen as setting off a "green revolution" or a "seed-fertilizer revolution" that would resolve the food-population problem that had been haunting this part of the world since the beginning of the sixties.[5] Indeed, the likelihood of some of the countries so heavily dependent on concessional food imports in the sixties, generating a sizeable exportable surplus of foodgrains appeared quite strong.[6] But

[2] To arrive at this conclusion, some authors additionally assumed declining fertilizer prices owing to technological advances in fertilizer industry. See, John Cownie, Bruce F. Johnston and Bart Duff, "The Quantitative Impact of the Seed-Fertilizer Revolution in West Pakistan: An Exploratory Study", *Food Research Institute Studies*, vol. IX, No. 1.

[3] The acreage estimates are for rice and wheat. See, Dalrymple, *op. cit.*

[4] Brown, *op. cit.*

[5] Opinion was divided whether the resolution of the food-population problem would be temporary or of a more durable nature.

[6] Pleasing though the prospect might appear to be, some observers became seriously concerned about the effects of demand limitations in the world market upon the growth of foodgrains output in these countries. Convinced of the strong possibility that foodgrains import demand would recede and that world prices would fall sharply in the next decade, they counselled these countries against measures to attain a rate of growth of foodgrains output higher than what could be attained through continued reliance upon their existing bullock-technology. See, Cownie, Johnston and Duff, "The Quantitative Impact," *op. cit.*

that was not all. It seemed idle to imagine that the effects of the green revolution could remain confined to food production alone; these seemed likely to be felt in other areas as well. The high yielding varieties are land-substituting, labour-using innovations; they are also neutral to scale and, therefore, usable by all farmers regardless of farm size. Consequently, they should raise income and employment of small farmers and farm labourers; they could "break the chains of rural poverty in important parts of the world."[7] Furthermore, these varieties require entirely new inputs and agronomic practices; their widespread use would raise the demand for modern (purchased) inputs like fertilizers and pesticides. In effect, therefore, the high yielding varieties appeared to be "engines of change" that would modernize and radically transform traditional agricultures; they would be, as one observer put it, "to the agrarian revolution of the poor countries what the steam engine was to the Industrial Revolution in Europe."[8]

At the same time, the possibility that the green revolution might easily turn out to be a Pandora's Box appeared very real. Misgivings related mainly to the social consequences of the green revolution: whether or not it would lead to a massive problem of equity and welfare. To begin with, how would the benefit from the green revolution be distributed between regions? On the whole, the distribution of benefits among regions would likely parallel existing resource endowment. The high yielding varieties would be the most productive in regions with favourable natural (agro-climatic) conditions and/or with substantial past investment in infra-structure development, such as irrigation, roads and power.[9] But these regions had already been in the forefront of agricultural progress and had already been prosperous in terms of per capita income from agriculture. Benefits to other regions with inadequately developed agricultural infra-structure and natural resources would be exceedingly small in comparison. The existing gap between the rich and the poor regions therefore

[7] Clifton R. Wharton, Jr., "The Green Revolution : Cornucopia or Pandora's Box?" *Foreign Affairs*, April, 1969.

[8] Brown, *op. cit.*

[9] John W. Mellor, "Report on Technological Advance in Indian Agriculture as it Relates to the Distribution of Income," mimeographed, December 12, 1969.

would increase and in this respect the green revolution could well exert a destabilizing influence.[10]

More importantly, the distribution issue that came to be of immediate concern was of an intraregional nature. One aspect of the issue is distribution of benefits among farm operators, large and small. Here, misgivings appeared to be warranted on *a priori* grounds. According to the generalized model of community adoption process, after a slow start, adoption of an innovation by farmers increases at an accelerated rate till about half the potential adopters come to adopt it; thereafter adoption increases but at a diminishing rate.[11] Viewed against time, the curve depicting percentage of adopting farmers rises very slowly in the initial stage, rapidly in the second stage and then tapers off. This is the familiar growth curve, or S-curve which implies that adoption of an innovation follows a normal distribution. Based on three segments of this distribution, adopters are generally classified into early adopters, majority and late adopters.[12]

Distinctive characteristics of the early adopters are that they are younger, more educated, venturesome and willing to take risks; they operate large farms and have high income and social status. Late adopters, in contrast, are older, less educated, security minded (conservative) operators of small farms with low income; they are complacent or sceptical and have lower social status.[13]

Given these characteristics, early adopters of an innovation would reap large income benefits and abnormal profits. By the time the majority comes to adopt the innovation, income gains realized by early adopters would disappear; unit costs of production would rise and product prices would fall. The average farmer, therefore, does not gain much from an innovation while the late adopters gain nothing.[14]

[10] Walter P. Falcon, "The Green Revolution : Generation of Problems," *American Journal of Agricultural Economics*, December 1970.

[11] See, Herbert F. Lionberger, *Adoption of New Ideas and Practices*, Iowa State University Press, Ames, Iowa, 1964.

[12] Rogers uses the two parameters of normal distribution—the mean and the standard deviation—to classify adopters into innovators, early adopters, early majority, late majority and laggards. See, Everett M. Rogers, *Diffusion of Innovation*, The Free Press of Glencoe, New York, 1962.

[13] Rogers, *ibid*; Lionberger, *op. cit.*

[14] See, Dana G. Dalrymple, *Technological Change in Agriculture : Effects and Implications for the Developing Nations*, U.S.D.A., April 1969.

In terms of this model of adoption process, it appeared that the 'first' or 'early' adopters of the high yielding varieties would be the wealthier, more progressive farmers. For them, as one observer put it, "it is easier to adopt the new higher-yield varieties, since the financial risk is less and they already have better managerial skills. When they do adopt them, the doubling and trebling of yields mean a corresponding increase in their incomes."[15] Owing to the differential rates of adoption, the rich farmers would become richer while the poor farmers would become poorer.

Further complications that could reinforce these tendencies could be easily anticipated. Along with yield and net returns, paid-out (or cash) costs of production for the high yielding varieties also turned out to be several times greater than the cash costs of production for the local varieties. To use the new varieties, therefore, the farmer must either have his own capital or have access to credit to finance his operations.[16] Poorer, smaller farmers being subject to resource constraints and without access to sources of credit, would be unable to take to the high yielding varieties even if they were willing to try them out.

Agrarian institutions in the underdeveloped countries could also come in the way of the smaller, poorer farmers adopting the high yielding varieties and receiving a share in the benefits from the green revolution. The credit agencies and the extension service, for instance, might serve only the larger, politically powerful farm operators;[17] the latter could easily preempt for their own use the bulk, if not the entire supply of scarce inputs like electricity, water and fertilizer. Consequently, the poorer farmers would be deprived of adequate inputs, so essential for successfully growing these new varieties.

Other aspects of the issue relate to distribution of benefits among owners and tenants, and among farmers and labourers. The advantages of the high yielding varieties appeared likely to flow in the direction of owners; rents were likely to rise. Owing to the increase in the profitability of farming and in returns to management, owners were likely to resume land for self-

[15] Wharton, Jr., *op. cit.*
[16] Wharton, Jr., *op. cit.*
[17] See, Mellor, *op. cit.*

cultivation.[18] Consequently, tenants would be thrown out of empoyment and forced to swell the rank of agricultural labourers. Thus, while the landowners would benefit from the green-revolution, the tenants would not.

As for landless farm labourers, while the high yielding varieties were likely to raise the demand for labour and employment per worker, the rise in real wages might be small. Furthermore, there was a strong likelihood that the green revolution would give an impetus to premature tractorization of operations and consequently would reduce employment of labour.[19]

On the whole, it appeared that the intraregional distribution of benefits from the green revolution would be highly inequitous. By and large, the adopters of the new technology would be the operators of large farms; while these operators were likely to make abnormal gains, the smaller, poorer farmers were likely to benefit little or not at all. Tenants and agricultural labourers would possibly be worse off. Consequently, existing disparities would increase and polarize the rural society. The resulting agrarian tensions and conflicts could easily lead to an explosive situation and from this point of view, the green revolution could well be a destabilizing factor in the underdeveloped world.[20]

Many of these expectations and misgivings, and much of the theorizing on the green revolution are based in some way or other, on the early experience with the high yielding varieties

[18] See, Mellor, *op. cit.*

[19] Bruce F. Johnston and John Cownie, "The Seed-Fertilizer Revolution and the Labor Force Absorption Problem", *American Economic Review*, September, 1969.

[20] Equity problems arising in the wake of the green revolution were characterized as "second generation problems" by Wharton. Recently, Falcon has drawn a distinction between the second and the "third generation problems". According to his classification, the first generation problems relate to the extension of the green revolution; the second generation problems include problems of marketing, storage, transportation, and resource allocation; the third generation problems concern equity, welfare and employment. See, Falcon, *op. cit.*

Attractive though these terms are, we hesitate to use them. Division of problems into proximate and distant may carry with it the implication that proximate ones are the most important while the distant ones are not. This implication is unfortunate and perhaps unintended by those who coined these terms. The problems of equity, welfare and employment are surely just as important and pressing as those relating to the diffusion of the new technology and both require immediate attention.

in South Asia, and in particular, in India. A large part of the now voluminous and still burgeoning literature on the subject continues to draw heavily upon the Indian experience either to generalize or to substantiate propositions concerning the effects of the green revolution. The reasons for this interest in the Indian experience are not difficult to explain. India has been one of the first countries to take to the high yielding varieties on a large scale as part of a "new agricultural strategy"; although the high yielding wheat varieties appeared simultaneously in India and Pakistan, India's experience with high yielding rice varieties is slightly older.[21] Besides, India accounted for between 65 and 70 per cent of the area planted under high yielding varieties in Asia in the early years of the green revolution; it is in India that the annual growth of area under high yielding varieties has been spectacular—from a mere 25,000 acres in 1965-66 to 18 million acres in 1970-71 (which was about 65 per cent of all areas under high yielding varieties in Asia that year). If all eyes were focussed on India, it was also because of the early recognition that the breakthrough in foodgrains production in India was an accomplished fact and that agriculture had begun to be transformed at last from a way of life to a modern industry.[22]

It was also in India that an equity problem of a staggering magnitude was seen to have already emerged in the countryside even in the early phase of the green revolution. Early observations, based on field trips and interviews of selected individuals, farmers, officials and traders, appeared in general to confirm the *a priori* expectations and apprehensions: that the first adopters had been the large and rich farmers who commanded irrigation and other resources and who could invest in machinery and equipment necessary for multiple cropping;[23] that limitations of capital had come in the way of adoption of high yielding varieties by farmers with 10 to 15 acres of land; that where the latter had been able to secure credit for investment in irrigation and machinery, they gained little, since debt repayment in succeeding years siphoned off a substantial part of their incomes; that small farmers with less than 10 acres of land had been wholly unable to take

---

[21]See, Dalrymple, *op. cit.*
[22]See, Brown, *op. cit.*
[23]Francine Frankel, *India's Green Revolution : Economic Gains and Political Costs*, Oxford University Press, Bombay, 1971.

advantage of the high yielding varieties because they had been unable to invest in the indivisible inputs—water and machinery.[24] The new agricultural strategy, for all its achievements, appeared also to be "the indirect cause of the widening of the gap between the rich and the poor."[25] Even in Punjab, the show-piece of the green revolution, early field trips suggested no more than ten per cent of the small farmers participating in the green revolution and, furthermore, the real sharing in the progress was restricted to relatively few, "perhaps only 10 and surely not more than 20 per cent of the farm households."[26]

Field trips also gave the impression that rents had been rising and that conditions were being created for a largescale eviction of tenants;[27] that the green revolution had hastened the pace of mechanization of farm operations and that a decline in employment of farm labourers could not be far behind.[28] The polarization of the rural society into the rich and the poor had begun and if the trend continued unabated it was feared that the green revolution might not remain green.[29] One incident occurring in Thanjavur district of Tamil Nadu state in December 1968 between landlords and workers resulting in the reported death of 43 persons was looked upon as a shadow of things to come;[30] an agrarian revolution was taken to be round the corner.

Eight years are a fairly long time in which to judge the performance of the green revolution and to evaluate its effects on output, income and employment.[31] From the evidence that is now

[24]*Ibid.*

[25]Wolf Ladejinsky, "The Green Revolution in Punjab : A Field Trip," *Economic and Political Weekly*, (hereafter referred to as *E&PW*) June 28, 1969.

[26]*Ibid.*

[27]Mellor, *op. cit.*; also Ladejinsky, *op. cit.*

[28]Ladejinsky, *op. cit.*

[29]Frankel, *op. cit.* Speaking of Punjab, Ladejinsky wrote that "the growing polarisation in agriculture between the rich and the poor is also an integral part of the Punjab scene." Ladejinsky, *op. cit.*

[30]Brown, *op. cit.*; Wharton, Jr., *op. cit.*

[31]From a broader scientific viewpoint, a period of eight years is admittedly too short to evaluate varietal research. The International Rice Research Institute (IRRI) was set up only in 1962 by the Rockefeller and the Ford Foundations; the International Corn and Wheat Improvement Centre (CIMMYT), also sponsored by the two Foundations did not get started till 1963, although wheat research had an early start. The Rockefeller Founda-

available, it should be possible to place the green revolution in its proper perspective. The purpose of this book is to do so but only in the context of one country, namely, India. Given the importance of the Indian experience and its role in the theorizing on the green revolution, this choice of the context would seem to be amply justified. But there is yet another reason for restricting the scope of this study to India alone, and it springs from a conviction that generalizations that ignore the differences in production environment obtaining in different countries are unlikely to be meaningful. The underdeveloped world in which the green revolution has made its debut is a heterogeneous collection of countries with varying degrees of infra-structure development. Even in respect of agrarian structure and institutions there is little in common among them. Since the production and equity effects of the green revolution depend upon these conditions, generalizations for the countries comprising the under-developed world would be of little use and could well be misleading. The effects of the green revolution need to be studied in a particular context and in a particular setting; and there may be no better setting, no better context than those provided by Indian agriculture.

The plan of this book is as follows. The growth of foodgrains output in India is analyzed in the next chapter to evaluate the performance of the green revolution and to place it in its proper perspective; alongside, the existing environment of agricultural production is compared with the technical characteristics of the high yielding varieties, with a view to determining the present and the future scope of the green revolution in India. The results of this evaluation provide the background against which discussion in the subsequent chapters is organized. Chapter three is devoted to one aspect of the issue of intraregional distribution of benefits from the green revolution, namely, the adoption pattern; based on the technical properties of the new varieties and the prevailing agronomic conditions of foodgrains production in India, a countrywide pattern of adoption is first hypothesized;

tion's wheat research programme in Mexico began in 1943. See, Brown, *op. cit.*; and Dalrymple, *Imports and Plantings, op. cit.* The concern of this book, however, is not with the progress of varietal research, but rather with the performance of the varieties that have been developed and put into the field in the last eight years.

the evidence on the adoption by farmers that is now available from various regions and from various sources is then examined. Chapter four deals with some issues in income distribution and particularly those concerning the relationship between land and income. The employment potential of the green revolution is assessed in chapter five and some of the issues involved in the controversy on the effect of tractorization on employment are delineated.

No assessment of the effects of the green revolution on the problems of equity and welfare can ever be complete without taking a look at some of the measures that are so often recommended as the solutions of these problems. One of these measures that has been strongly urged in the context of the green revolution—a measure that is in fact the conventional answer to the equity and welfare problems in agriculture—is land reform. Chapter six examines the relevance and adequacy of this measure to resolve particularly the problem of unemployment or labour absorption in agriculture. Finally, chapter seven turns to the broader issues beyond the green revolution—those relating to growth both in the farm and the nonfarm sectors.

Chapter Two

THE EXTENT OF THE GREEN REVOLUTION

Viewed in the context of India's rapidly rising output of food-grains between 1967-68 and 1970-71 it is not surprising that the green revolution initially generated an unbridled optimism about food prospects for the growing millions. Aggregate foodgrains output had reached the level of 88.9 million metric tons in 1964-65 —a year of unusually favourable rainfall. The new agricultural strategy, comprising high yielding varieties, fertilizers and pesticides was initiated on a modest scale in 1965-66,[1] but its immediate effects were obscured by the unprecedented drought conditions prevailing in 1965-66 and 1966-67. Foodgrains output in 1967-68 totalled 95 million metric tons and in 1970-71 about 108.4 million metric tons. Sustained increase in foodgrains output of this magnitude appeared to many as deserving the label of a "revolution" and to amply justify some of the early hopes reposed in the green revolution.

One way of evaluating the recent increases in annual foodgrains output is to compare them with the long run trend rate of growth established in the period 1949-50 to 1964-65, that is the period before the advent of the high yielding varieties. If output increases turn out to be substantially higher than what would be expected on the basis of the past trend, one could conclude that the green revolution has had an impact on foodgrains output. The expecta-

[1] Hybridization of selected crops (maize and millets) began in India in 1960. Mexican dwarf varieties of wheat were tried out on a selected basis in 1963-64. Exotic varieties of paddy seeds, such as Taichung Native 1, were introduced in 1965. However, the new strategy became fully operational only from *kharif* 1966, when propagation of high yielding varieties was taken up as a full-fledged programme over a large area. See, Government of India, Planning Commission, *Fourth Five Year Plan*, p. 113.

tion is that since the green revolution represents a sharp break from the past, it will have raised the long run trend rate of growth of foodgrains output considerably over the rate of 3.05 per cent per annum established during the period 1949-50 to 1964-65.

In the upper part of Figure 2.1, the index numbers of foodgrains output (in natural logarithm) have been plotted against time; the reason for using index numbers rather than absolute figures is that the former are supposed to be more reliable and comparable than the latter.[2] The long run trend of foodgrains production (1949-50 to 1964-65) has been sketched in on the graph as a broken line. Foodgrains output would have continued to grow at this rate had there been no green revolution. On the graph, a new trend line for the years 1949-50 to 1970-71 (excluding the drought years of 1965-66 and 1966-67)[3] appears as an unbroken straight line. Comparison of the two trend lines clearly shows that there has been some improvement in the rate of growth of foodgrains production—but an improvement that can be termed at best nominal. It appears that in terms of rates of growth of foodgrains output, the achievement of the green revolution has been grossly overrated. The long run trend growth rate improves from 3.05 to only 3.26 per cent per annum when the output increases between 1967-68 and 1970-71 are taken into account. It is hard to take this small improvement as being significant enough to merit the term revolution.

The record in respect of foodgrains yield is no different. The lower part of Figure 2.1 shows the improvement that took place in foodgrains yield as a result of the green revolution. The long run trend rate of growth in yield rose from 1.63 in 1949-50 to 1964-65 to 1.99 per cent per annum in 1949-50 to 1970-71.[4]

The trend rates of growth depend significantly on the choice of the base and terminal years. For instance, the improvement that appears in Figure 2.1 would be considerably reduced if the terminal year was taken to be 1969-70 instead of 1970-71. Like-

[2] See, *Estimates of Area and Production of Principal Crops in India*, 1970-71, p. 151.

[3] B. S. Minhas and T. N. Srinivasan have argued that the years 1965-66 and 1966-67 were abnormal years and should be excluded from trend fitting. See, "Food Production Trends and Buffer Stock Policy," *The Statesman*, November 14-15, 1968.

[4] Data for 1968-69 through 1970-71 are still provisional and are likely to undergo revisions at a later date.

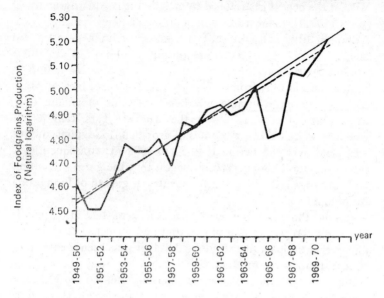

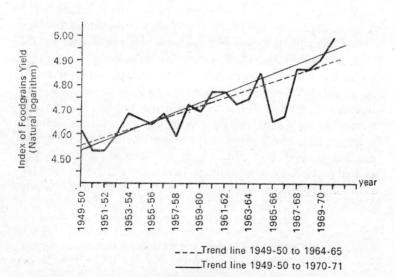

- - - - Trend line 1949-50 to 1964-65
———— Trend line 1949-50 to 1970-71

FIGURE 2.1

wise, if the period is stretched to include the last two agricultural years (1971-73), the record may show no improvement at all. Although firm data are not available, preliminary estimates suggest a decline in foodgrains output in 1971-72 of about 3.3 million metric tons and a decline in 1972-73 of at least 10 to 13 million metric tons from 1970-71 levels. These declines, attributable to the unfavourable weather conditions prevailing over large parts of the country in the last two years, cannot but drastically scale down the rates of growth of foodgrains output and yield over the period 1949-50 to 1972-73; consequently, a later evaluation may very well suggest that the green revolution had not raised the long run trend rates of growth of foodgrains output and yield at all.

It is one thing to argue, as we have done here, that the improvement in aggregate foodgrains output and yield has been small; and it is entirely another to deny that there has been a change in Indian foodgrains production.[5] There have always been a few sceptics to whom the green revolution has been no more than a myth and who have always been eager to attribute the increase in output witnessed since 1967-68 to favourable rainfall. The production setbacks in the last two years may seem to lend support to their position.

It should be conceded that weather conditions indeed had a hand in the increase of foodgrains output in the period between 1968-69 and 1970-71, just as these have been responsible for the decline of output since 1970-71. In fact, weather will continue to be an important factor in deciding the level of output notwithstanding the technological advances that have taken place, and output will continue to fluctuate from year to year. It cannot be otherwise, since about 80 per cent of the area under foodgrains is unirrigated, and since a large proportion of this area lies in drought-prone districts with highly uncertain and extremely low annual precipitation; in addition, (and we shall have more to say about this point later on) a large proportion of the area officially

---

[5] A former chairman of the Agricultural Prices Commission even went to the extent of claiming that the growth rate of foodgrains output had actually declined in the sixties. See, Ashok Mitra, "Bumper Harvest has Created Some Dangerous Illusions," *The Statesman*, October 14-15, 1968. Minhas and Srinivasan (*op. cit.*) have shown that such claims are based on adoption of inadmissible procedures in the fitting of trends.

classified as irrigated, is no better than unirrigated land anyway, dependent as it is on rainfall as the source of water.

With this important caveat entered, one can safely assert that there is no evidence to show that a change has not occurred in Indian foodgrains production; the evidence does not show that increases in output obtained since 1967-68 are attributable exclusively to favourable rainfall. In fact, even when the effect of weather is held constant in statistical analyses, the effects of the new technology show up quite strongly in the form of enhanced output.[6] The question, therefore, is not whether a change has occurred in Indian agriculture, but whether the change is extensive and far reaching.

A useful way to approach this question is to look at the components of the foodgrains basket separately and identify the crops that have recorded substantially improved rates of output growth. In Table 2.1, the compound rates of growth of output and yield of the principal foodgrain crops are shown separately for the periods 1949-50 to 1964-65 and 1949-50 to 1970-71.[7] The actual annual output and yield for these crops are also shown in Figure 2.2 through Figure 2.6. It will be observed that considerable improvement has indeed taken place in respect of output and yield of bajra, maize and, of course, wheat. The long run trend rates of growth of output and yield of bajra have risen from 2.35 and 1.23 per cent per annum respectively to 3.53 and 2.23 per cent per annum; those of maize from 3.85 and 1.17 to 4.54 and 1.31 per cent; and those of wheat from 3.99 and 1.27 to 5.87 and 2.80 per cent per annum. Of the three crops, the growth of output and yield of wheat alone has been what may be called spectacular. There is some truth in the cynical state-

[6] See, Ralph W. Cummings, Jr., and S. K. Ray, "The New Agricultural Strategy: Its Contribution to 1967-68 Production," *E&PW* March 29, 1969; and "1968-69 Foodgrains Production: Relative Contribution of Weather and New Technology," *E&PW*, September 27, 1969. It is significant that preliminary estimate of foodgrains production (released after the manuscript went to press) of 95 million tonnes in 1972-73 despite widespread drought conditions, is more than 22/23 million tonnes higher than the output level attained in the drought years of 1965-66 and 1966-67. It is arguable if this level of production could be attained without the use of the high yielding varieties.

[7] For reasons advanced earlier, the growth rates have been worked out by excluding the years 1965-66 and 1966-67.

TABLE 2.1 : GROWTH RATES OF OUTPUT AND YIELD OF
SELECTED FOODGRAINS, INDIA

Crops			Period	Output	Yield
All foodgrains	...	...	1949/50—1964/65	3.05	1.63
			1949/50—1970/71	3.26	1.99
Rice	...	...	1949/50—1964/65	3.47	2.12
			1949/50—1970/71	3.26	1.94
Jowar	...	...	1949/50—1964/65	2.50	1.49
			1949/50—1970/71	1.94	1.18
Bajra	...	...	1949/50—1964/65	2.35	1.23
			1949/50—1970/71	3.53	2.23
Maize	...	...	1949/50—1964/65	3.85	1.17
			1949/50—1970/71	4.54	1.31
Wheat	...	...	1949/50—1964/65	3.99	1.27
			1949/50—1970/71	5.87	2.80
Pulses	...	...	1949/50—1964/65	1.64	—0.24
			1949/50—1970/71	0.95	0.04

ment that the green revolution in effect has been a wheat revolution.

The growth rates of output and yield of jowar and rice in Table 2.1 do not suggest any improvement; in fact, the growth rates are distinctly lower. Thus the long run trend rate of growth of output and yield of jowar has declined from 2.50 and 1.49 respectively in the period 1949-50 to 1964-65 to 1.94 and 1.18 per cent per annum in the period 1949-50 to 1970-71. Similarly, in the case of rice, the growth rates of output and yield have fallen from 3.47 and 2.12 in 1949-50 to 1964-65 to 3.26 and 1.94 respectively in 1949-50 to 1970-71. It appears that the green revolution is yet to touch these two crops.

On the whole, the effects of the green revolution on the output and yield of foodgrain crops have been mixed: spectacular for wheat, some improvement for bajra and maize, and no improvement at all for jowar and rice.[8] In the foodgrains output index wheat has a weight of 12.6 per cent only; the combined weight of wheat, bajra and maize is no more than 19.8 per cent. In contrast, rice has a weight of 52.7 per cent and together with jowar

[8] There has been no breakthrough in regard to pulses production; the negative growth rate of pulses output in the period 1949-50 to 1970-71 should be a cause of concern in view of the fact that pulses provide a major source of protein for the bulk of India's population.

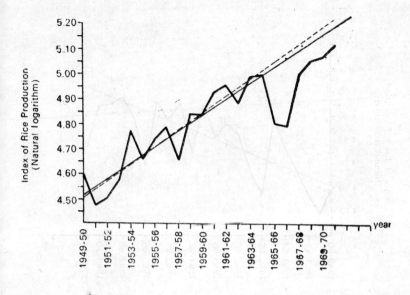

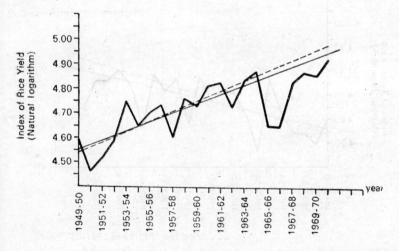

----Trend line 1949-50 to 1964-65
——Trend line 1949-50 to 1970-71

FIGURE 2.2

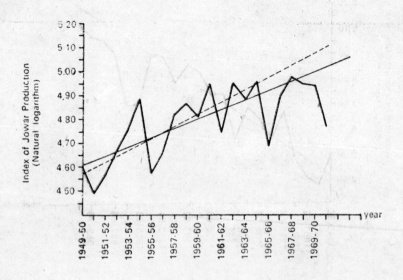

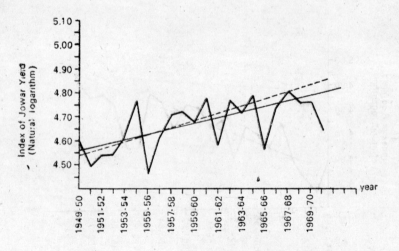

_ _ _ _ Trend line 1949-50 to 1964-65

Trend line 1949-50 to 1970-71

FIGURE 2.3

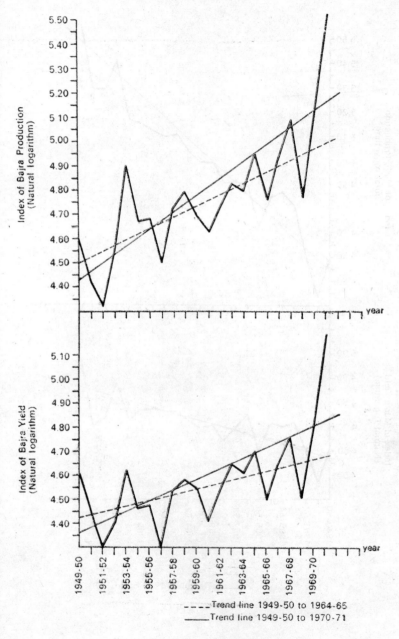

FIGURE 2.4

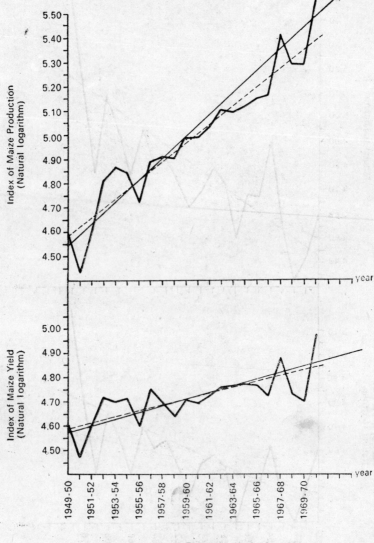

FIGURE 2.5

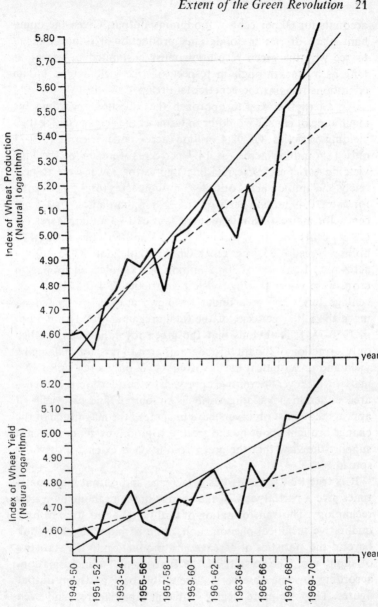

FIGURE 2.6

accounts for 60 per cent of foodgrains output. Given the dominant place of rice in foodgrains production it is not difficult to see why the green revolution must be limited in extent as long as a breakthrough in respect of rice yield under Indian conditions remains the geneticists' dream.

An alternative way to approach the question regarding the magnitude of change would be in terms of the acreage under high yielding varieties. Of 92.4 million acres under rice in 1970-71 only 13.6 million acres (or 14.7 per cent) were under the high yielding varieties; corresponding figures for jowar were respectively 1.97 million acres out of 43 million acres (or 4.5 per cent); for bajra, 0.49 million acres out of 31.8 million acres (or 1.5 per cent); for maize, 0.98 million acres out of 14.3 million acres (or 6.9 per cent) and for wheat about 15.81 million acres out of 44.2 million acres (or 35.7 per cent). On the whole, only 32.85 million acres out of an area of 306.2 million acres under all foodgrain crops were under the high yielding varieties in 1970-71; in percentage terms the area under the new varieties comprised no more than 10.7 per cent of the total area under foodgrain crops in 1970-71. It is obvious that the green revolution has touched only a fraction of the total foodgrains area so far; and judging by yield and/or output, it has involved as yet only a few crops. No matter whichever alternative approach to the question is chosen— area under high yielding varieties or output and/or yield—it appears that most observers have misjudged the magnitude of the change brought about by the green revolution; evidence does not support the view that the green revolution is extensive or widespread.

It is true that the statistics of acreage and output may sometimes give a partial view of the change brought about by a new technology. The transformation of traditional agriculture being a qualitative technical phenomenon,[9] one perhaps ought to look beyond the statistics of acreage and production to the statistics of input use. The results of the fundamental transformation, according to some observers, "are visible only partly in output figures. They are more clearly visible in input figures and even

[9] M. L. Dantwala, "From Stagnation to Growth: Relative Roles of Technology, Economic Policy and Agrarian Institutions," Presidential Address, Indian Economic Association, Gauhati, December 1970.

more in the behaviour patterns of today's farmers."[10]

Once again it is necessary to assert that what is in question is not whether a change has occurred in Indian agriculture, but whether the change is total and extensive. The statistics of input use as well as the statistics of acreage and production certainly testify to the fact that a change has occurred, but they do not suggest that the change is extensive. Briefly, the highlights of input use are as follows. The consumption of fertilizers in terms of nutrients almost doubled between 1967-68 and 1970-71, and so did the number of tractors; the number of tubewells installed between 1966-67 and 1970-71 was around 275 thousand and the annual increase in the number of energized pumpsets is reported to be about two hundred thousand in the same period.

While this increase in input use is welcome, one ought to be cautious in accepting this increase as evidence of an input revolution or of a fundamental transformation of Indian agriculture. The fact remains that the level of use of modern inputs at the time the high yielding varieties appeared on the scene was dismally low; a large percentage change from such a low base tends to magnify the importance of this change. For instance, fertilizer consumption in 1964-65 was no more than six hundred thousand tonnes or barely 4 kilograms per hectare in terms of nutrients; the increase in fertilizer use since then has raised this figure to about two million tonnes (1970-71) or about 12 kilograms per hectare.

A careful perusal of the data on fertilizer consumption would reveal furthermore that the upsurge in the use of fertilizer has been confined to a few states. Consumption of fertilizer per hectare of agricultural land in 1970-71 did not exceed 2.50 kg in five states and Union Territories (Assam, Manipur, Tripura, Nagaland and Himachal Pradesh); 3.50 kg in three states (Madhya Pradesh, Rajasthan and Orissa); 10 kg in five states (Bihar, West Bengal, Jammu and Kashmir, Maharashtra, and Goa). Conspicuous in fertilizer use per hectare were one Union Territory and two states, namely, Pondicherry (117.04 kg), Punjab (40.24 kg) and Tamil Nadu (34.50 kg). In the rest of the states (Mysore, Gujarat, Haryana, Uttar Pradesh, Kerala, and Andhra Pradesh) fertilizer use varied from 12 kg to 19 kg per hectare.[11] These data

---

[10]A. M. Khusro, "Problems of the Green Revolution," *The Times of India*, New Delhi, August 15, 1972.

[11]*Fertiliser Statistics*, 1971-72, The Fertiliser Association of India, pp. 206-7.

seem to suggest that increase in fertilizer use has not been a general phenomenon. Although relevant data regarding other inputs are not available as yet, one may not be far wrong in expecting that the input revolution would be found to be limited in geographical extent.[12]

It is pertinent to ask, why did the initial expectations regarding output growth and pervasiveness of the green revolution go awry? The answer would seem to lie in the failure to appreciate the implications of the technical properties of the high yielding varieties, and the point is best illustrated with reference to the high yielding rice varieties. These varieties, unlike those of wheat need to be adapted to each region to fit local conditions of production. Some of the early varieties were susceptible to pests, bacterial leaf blight and gallmidge; some varieties were also temperature sensitive and unsuitable to the cool night temperature in North India; others, because of their non-photo period sensitiveness and early maturity had to be harvested in the midst of the rainy season.

In fact, a breakthrough in rice yield is hard to achieve under Indian conditions. Rice is grown in India mainly as a *kharif* crop during the monsoons, primarily in low-lying areas that easily become water-logged, providing conditions that are supposed to be ideal for the local varieties. When grown as a *kharif* crop, however, the yield potential of the high yielding rice varieties is seldom realized. Both water-logging and overhanging cloud cover during maturity considerably reduce their yield; additionally, they become highly susceptible to pest attacks under these conditions.[13] On the whole, the new rice varieties fare much better when grown as a *rabi* crop in irrigated, well-drained upland areas; strong sunshine alone in the *rabi* season adds about 20 to 50 per cent to the yields. Under these conditions, the high yielding

[12]There is evidence that the change in farmer behaviour and attitude too is confined to a few regions. We do not elaborate the point here, but interested readers would find the following articles by Ashok Thapar in *The Times of India* useful: "Changing Punjab Village," June 14, 1971; "New Challenge to Agriculture," August 15, 1972.

[13]See, Randolph Barker and Mahar Mangahas, "Environmental and Other Factors Influencing the Performance of New High Yielding Varieties of Wheat and Rice in Asia," in *Agricultural Development in Developing Countries— Comparative Experience*, Indian Society of Agricultural Economics, Bombay, October 1972.

potential of the new varieties can be achieved easily. But given the extremely small proportion of area under irrigated, upland *rabi* rice—that is, the area with favourable environmental conditions—the impact of the high yielding varieties on rice area, output and yield would continue to be small and limited.

This review of the technical attributes also brings out the most important technical property of the high yielding varieties which, in the light of subsequent events, appears to have been either overlooked or inadequately appreciated by most observers; this property is the dependence of the high yielding varieties on irrigation. Realization of the yield potential of the high yielding varieties depends critically on controlled application of water and on availability of drainage facilities. These seeds need water at specific periods of growth and the timing of irrigation is crucial for optimum yield and optimum returns to associated investment in inputs like fertilizers. In the case of wheat, for example, timing and spacing of irrigation can raise yield by as much as 40 per cent.[14] First irrigation at the time of crown root initiation, that is around the third week of planting alone raises yield by as much as 30 per cent than when it is delayed, as in the case of local varieties, to the sixth week.[15] Similar critical stages for irrigating the high yielding varieties of wheat are maximum tillering and the dough stages. On heavier soils, three more irrigations and on light soils, two additional irrigations may be necessary.[16] In fact, experiments suggest that faulty irrigation schedule can reduce the yield of high yielding varieties of wheat by about 29 per cent. For rice, lack of moisture in the primordial initiation, flowering and milk stages can reduce yield by as much as 50 per cent.[17] Clearly, to meet water requirement of this kind, there must be controlled sources of water and an adequate public irrigation network.

At this stage, one point needs to be emphasized. We are not suggesting that the high yielding varieties cannot be grown at all on unirrigated land. They can perhaps be grown in areas

[14]N. G. Dastane, "New Concepts in Irrigation: Necessary Changes for New Strategy," *E&PW*, March 29, 1969.

[15]J. S. Kanwar, "From Protective to Productive Irrigation," *E&PW*, March 29, 1969.

[16]R. G. Anderson, "Wheat Improvement and Production," The Rockefeller Foundation, New Delhi (mimeographed) September 13, 1968.

[17]Government of India, *Report of the Irrigation Commission*, Vol. 1, 1972.

supposedly blessed with "assured rainfall". But the area officially classified as under assured rainfall does not mean much since the adequacy of the total quantum of rainfall cannot be taken for granted; nor is the distribution of rainfall in these areas over the season such as to meet the exacting demands of the new varieties. Furthermore, water and humidity being uncontrollable, the realized yield of the new varieties may not significantly differ from that of the local varieties. Yet, one cannot entirely rule out the possibility of some unirrigated farms using the new varieties on what may be called an experimental basis, at least in the initial phase of the green revolution. In all likelihood, their number would be small and the unirrigated area thus planted under the high yielding varieties would be too trivial and insignificant to make any impact on the aggregate area under these varieties. In all likelihood, a large proportion of these farmers, having experienced the risks involved and the poor returns, would revert to the cultivation of local varieties before long.[18]

We are arguing that the complementarity of the new seeds with water effectively limits the domain of the green revolution. In fact, the high yielding varieties have a specific domain of relevance: they concern a fraction of cultivated area—the irrigated (foodgrains) land—and a subset of farmers—those who operate it. At least initially, these varieties would be restricted to those areas and to those farms that have presently adequate sources of irrigation. Elsewhere, only after the required investment has been made and an adequate irrigation system has been developed can the high yielding varieties become applicable and relevant. Failure to weigh this property of the new varieties against the production environment obtaining in Indian agriculture seems to account for the early euphoria over the pervasiveness and the far-reaching consequences of the green revolution.

To be sure, there are a number of other inputs which are required for the cultivation of the high yielding varieties: fertilizers, pesticides, and above all, the operating capital or cash for the purchase of these inputs; to this list of requirements one could also add a package of cultural practices. Yet, none of these is as important as the availability of irrigation which alone is the key prerequisite for determining the domain of the new technology,

[18]This seems to be the most important factor responsible for the phenomenon of high rate of drop-outs from the HYVP that has puzzled some analysts.

or of its applicability on a given area or farm. For instance, lack of operating capital, or of credit may at the most restrict the application of fertilizer, and therefore, may at the most preclude the realization of the maximum potential yield; it does not, however, prevent an operator, if he has an irrigated holding, from growing the high yielding varieties. The yields of the new varieties being considerably greater than those of the local varieties even at zero level of fertilizer application, the operator of an irrigated holding can always harvest a larger crop even if he has no cash to purchase fertilizer. The operator of an unirrigated holding is not so fortunate however; the absence of irrigation facilities entirely rules out the possibility of growing the high yielding varieties. Without irrigation, a surfeit of other inputs would be of no avail.

How extensive the green revolution can ever be may be judged in terms of the total available irrigated area under foodgrains. Out of India's 297.4 million acres under foodgrains in 1969-70, only about 63.7 million acres were under irrigation and this is the area that is relevant to the new varieties. The entire irrigated foodgrains area is scheduled to be covered by the high yielding varieties by the end of 1973-74. When this target is fulfilled, a little more than 1/5th of the total foodgrains area would be covered by the green revolution.

This too is an optimistic estimate. For, a large proportion of the area officially classified as irrigated is no better than unirrigated land, depending on rainfall as the source of water. In many canal-irrigated districts, such as Raipur in Madhya Pradesh, the network was primarily designed to supplement rainfall in the months of August and September; the canals are dry the rest of the year.[19] For the most part, the networks built in the past were designed to provide protection against drought rather than productive irrigation.

Many of the irrigation systems date back to the nineteenth century or earlier, and these cannot be expected to meet the exacting demands of the high yielding varieties for water. The usefulness of these systems is severely limited, as the Irrigation

[19] *Modernizing Indian Agriculture*, Report on the Intensive Agricultural District Programme (1960-68), vol. I, Expert Committee on Assessment and Evaluation, Ministry of Food, Agriculture, Community Development and Cooperation, New Delhi, 1969.

Commission observes, "by structural handicaps, such as outmoded headworks, the absence of suitable silt-excluding devices and unsatisfactory arrangements for cross drainage. These handicaps are reinforced by other factors such as faulty irrigation management and poor drainage and distribution."[20] Even some of the comparatively recent systems, such as the Sarda Canal System, the largest canal network in Uttar Pradesh, are no better either. As the Irrigation Commission points out, the inadequacies of the system come in the way of farmers taking to the cultivation of high yielding wheat varieties in its command area.

In several areas, as in Aligarh, the canal irrigation system is not in a position to meet the water requirement of the high yielding varieties "as the available water has been obligated to thousands of farms relatively thinly for meeting the requirements of traditional varieties."[21] In North India, a cusec of water that could meet the requirement of wheat during the stages of heading and grain development on only 60 acres is supplied thinly over 200 to 400 acres by the tubewell and canal systems operated by government irrigation departments.[22] Even in districts like Thanjavur in Tamil Nadu, and in West Godavari in Andhra Pradesh—that is, districts which are supposedly blessed with well-developed irrigation systems—irrigation water is not controllable at the farm level and, therefore, is not capable of being put fully to productive use.[23] In canal-irrigated rice growing areas the common practice "is to flow water from field to field, down and across a terraced area from the supply channel at the top to a drain at the terrace base. It may take as much as a month for water to flow across the terrace and cover all the fields from top to bottom."[24] Since water cannot be controlled at the field level, such areas may be only marginally relevant to the high yielding varieties.[25]

[20] *Report of the Irrigation Commission, op. cit.*

[21] *Modernizing Indian Agriculture, op. cit.*

[22] W. David Hopper, "The Mainsprings of Agricultural Growth," Dr. Rajendra Prasad Memorial Lecture to the 18th Annual Conference of the Indian Society of Agricultural Statistics, January, 28-30, 1965.

[23] *Modernizing Indian Agriculture, op. cit.*

[24] See, W. David Hopper and Wayne H. Freeman, "From Unsteady Infancy to Vigorous Adolescence: Rice Development," *E&PW*, March 29, 1969.

[25] Drainage is equally important. Rice yields are low more because of water-

The scope of the high yielding varieties could be enlarged to a certain extent by extending irrigation facilities to unirrigated areas. But the additions to the irrigated area, it needs to be clearly understood, would be far from spectacular and dramatic; in all likelihood, it would be a painfully slow growth over time. The scope for a rapid extension of irrigation is severely limited. A large part of the more easily available water resources has already been exploited.[26] Areas with proven resources of water than can still be tapped with relative ease are few. It is also clear that much of the new investment for irrigation has to come from the public agencies; few farmers have the requisite resources to develop irrigation and in any event farmer investment in irrigation is unlikely to result in a considerable addition to the existing area under irrigation.

Since most of the easily accessible water resources have been already developed, it follows that further expansion of irrigated area would be slow and extremely costly. Cost of developing irrigation at the time of the first Five Year Plan was estimated at round Rs. 500 per acre. Current costs are about three times that figure and future projects are likely to be more expensive because of difficult locations and complex technology involved.[27] Some observers continually dwell on the theme of a low cost route to rapid agricultural development now made possible by the high yielding varieties.[28] They seem to ignore the fact that a massive investment in irrigation, among other things, is a precondition for rapid growth of agriculture. Indeed, there is no easy, inexpensive route to agricultural development.[29]

---

logging and excess water than due to lack of water. See, M. S. Swaminathan, "Scientific Implications of HYV Programme," *E &PW* Annual Number, January, 1969.

[26] See, K. N. Raj, "Some Questions Concerning Growth, Transformation and Planning of Agriculture in the Developing Countries," in E. A. G. Robinson (ed) *Economic Development in South Asia*, Proceedings of a Conference held by the International Economic Association at Kandy, Ceylon, pp. 1-12, 1969.

[27] *Report of the Irrigation Commission, op. cit.*

[28] See, Bruce F. Johnston and John Cownie, "The Seed-Fertilizer Revolution and the Labor Force Absorption Problem," *American Economic Review*, September 1969.

[29] Examining the growth of rice production in the Philippines, Thailand and Taiwan, Hsieh and Ruttan reach the following conclusion for countries in South and South East Asia: "The magnitude of the investment required to

The scope of the green revolution could also improve owing to the development of better varieties suitable for use under *kharif*, rainfed conditions. While the possibility of developing such varieties cannot be denied, it will be unreasonable to entertain hopes of a spectacular early breakthrough. If anything, the progress in this direction is likely to be slow. To take one example, hybrid corn varieties were released for commercial use in the United States in the early 1930s; these varieties turned out to be suitable for the central cornbelt but not for the southern corn growing areas. It took some 20 to 25 years before varieties suitable for the latter region could be developed. The varieties that are about to be released in India, or that are in advanced stages of trial are, almost without exception, meant for cultivation under dry, irrigated conditions. Relatively little attention has been given so far to the development of varieties suitable for *kharif* cultivation. The varieties that could raise the production of rice, jowar and bajra substantially under rainfed conditions, therefore, will be in the making for a long time to come. It will be unrealistic to expect otherwise, despite optimistic reports about the impending release of a new line of superior varieties, particularly for rice.

In sum, the extent of the green revolution in India has been small, both in terms of area covered and impact on output. The future scope too is severely limited. Annual additions to the area planted with high yielding varieties would be slow after 1973-74 and would be determined mainly by the annual additions to irrigated area under foodgrains. Growth rate of the area under the new varieties as well as growth rate of output of foodgrains would be far from spectacular and would likely taper off

---

realize the production potential inherent in the new technology that is being created tends to be substantially underestimated. There will have to be massive investment in the industries that produce the inputs of fertilizer and insecticides; there will have to be massive investment in irrigation if the investment devoted to development of new varieties and production of the technical inputs is to achieve a reasonably high return; and it will be necessary to commit substantial increases in trained manpower to the tasks of management related to the direct investment and to educational work associated with rapid achievement of the production potentials." S. C. Hsieh and V. W. Ruttan, "Environmental, Technological and Institutional Factors in the Growth of Rice Production : Philippines, Thailand and Taiwan," *Food Research Institute Studies,* vol. 7, No. 3, 1967.

once the currently irrigated foodgrains land is wholly covered by the high yielding varieties.

It is against this background that the fancied effects of the green revolution on interfarm disparities, income and employment need to be viewed. In retrospect, expectations concerning the resolution of the problems of rural unemployment and poverty, as well as predictions of an emergence of a massive problem of equity and welfare, appear to have been based on an insufficient appreciation of the scope of the green revolution on the one hand, and the realities of Indian agriculture on the other. In the following chapter we turn to one aspect of the question of distribution of benefits from the green revolution, namely, the pattern of adoption.

Chapter Three

THE ADOPTION PATTERN

In the literature on the green revolution, the issue of intraregional distribution of benefits has centred on the question as to whether the adoption of the new technology, represented by the high yielding varieties of seeds, fertilizers and pesticides, would be confined to a relatively small number of operators of *large* farms to the exclusion of the more numerous, less affluent operators of *small* farms.[1] The issue has been posed, it should be noted, in a narrow "small-vs-large farm" framework—one that completely disregards the existence of farms that are neither large nor small. It is true that in terms of the generalized model of adoption, as noted in chapter one, there are reasons for expecting different rates of adoption on these two groups of farms; the reasons specifically cited in the literature on the green revolution have ranged from difference in ability to bear risks[2] to differences in political power enjoyed by these two groups,[3] from resource constraints on small farms to the initial diversity of income and differences in absolute and marginal spending behaviour at different income levels.[4] Though, at first sight, these reasons appear to be plausible, restriction of the field of enquiry only

[1] See, William J. Staub and Melvin G. Blase, "Genetic Technology and Agricultural Development," *Science*, July 9, 1971; also, Francine Frankel, *India's Green Revolution*, *op. cit.*

[2] Clifton R. Wharton, Jr. "The Green Revolution : Cornucopia or Pandora's Box?" *Foreign Affairs*, April 1969.

[3] Cf. "As compared to smaller cultivators, the larger farmers can better afford the risks of innovation and they wield more political power over the development agencies which provide access to credit and crucial supplies such as fertilizer, seed and pesticides." Uma J. Lele and John W. Mellor, "Jobs, Poverty and the "Green Revolution," *International Affairs*, January 1972.

[4] Uma K. Srivastava, Robert W. Crown and Earl O. Heady, "The 'Green Revolution' and Farm Income Distribution in India: A Theoretical Analysis," *E&PW*, December 25, 1971.

to these categories of farms is apt to lead to a restricted view of the adoption pattern itself and perhaps to erroneous conclusions. The discussion in the preceding chapter concerning the scope of the green revolution and the domain of relevance of the high yielding varieties provides, in our view, the clue to the pattern of adoption that is likely to emerge in the Indian context. Since irrigation is the "key prerequisite for determining the applicability of high yield varieties on a given farm,"[5] the adoption pattern would be determined mainly by the distribution of irrigated farms in different size-classes; furthermore, the scope for a rapid extension of irrigation being severely limited, annual additions to the body of irrigated farms are unlikely to change significantly the existing distribution of irrigated farms. There is, therefore, no escape from the fact that the adopters of the high yielding varieties initially would be farms that are already irrigated; for a long time to come these farms would also constitute the bulk if not the entire set of adopter farms.

It will be readily conceded that irrigation, though the key prerequisite, is not the only factor determining the adoption of high yielding varieties; while irrigation would determine the "potential" pattern, the realized pattern of adoption is likely to be influenced somewhat by factors like the availability of critical inputs such as quality seeds, fertilizers and pesticides, and the ability of farmer with irrigated holdings to purchase these inputs. The restricted supply of these inputs and the lack of ability of farmers to purchase them pose the problem of resource constraints, and it will be examined in some detail in this chapter later on.

The expectation that the adoption pattern would closely follow the distribution of irrigated farms without being unduly impeded by constraints on resources, is based also on the hitherto neglected fact that the adoption of the high yielding varieties in India has been taking place under the influence of a government initiated mass action programme, called the High Yielding Varieties Programme (HYVP). In the HYVP, credit is channelled through the cooperatives, foundation and certified seeds are supplied by the state governments and the National Seed Corporation, while fertilizer supply is arranged by the central government. Initiated in 1965, the HYVP comprises the following

[5] Staub and Blase, *op. cit.*

phases. First, an area is selected where a large proportion of cultivated area is under assured irrigation; second, a time-table is prepared for completion of specific tasks involved and the input and credit requirement of participants is worked out; third, arrangement is made for the supply and distribution of inputs, field staff is strengthened and coordination established between concerned government departments; finally, participants are selected and exposed to trials and demonstrations. The criteria for the selection of farmer participants are: (a) that they have assured irrigation; (b) that they are members of cooperative credit societies; and if they are not, (c) that they are willing to become members of cooperatives and to invest in the inputs.[6]

The goal of the High Yielding Varieties Programme is to cover the entire irrigated land under foodgrains before the end of the Fourth Plan period, and to do so, the programme must involve all irrigated farms irrespective of size. The adoption issue, therefore, seems to call for an examination of the distribution of irrigated farms to determine whether it is biased in favour of or against a specific size-class of farms.

Distribution of Irrigated Farms

Although the issue has been posed in a regional context, there can be no one regional pattern, since no two regions, however defined, are exactly alike in respect of distribution of irrigated land among farms. Consequently, the adoption pattern too would likely vary from location to location and indeed from village to village.[7] The quest for generality in the face of this regional diversity is, therefore, likely to be fruitless. Even so, a study of the distribution of irrigated land and of the adoption pattern would be rewarding in itself, say, at the level of the village or that of the block. However, such a study must wait till micro level data are available. At present data on irrigation distribution are not available even at the level of the states.

Under the circumstances, a reasonable procedure would be

---

[6] See, *Planning and Implementation in Agriculture : Studies on High Yielding Varieties Programme*, vols. I and II, Indian Institute of Management, Ahmedabad, 1967.

[7] This would be evident from the results of the empirical investigations presented in the latter part of this chapter.

to examine the countrywide distribution of irrigated land by farm size and to recast the discussion in an all-India perspective. This procedure would at least enable us to judge if for the country as a whole, the adoption pattern is likely to be biased towards a specific category of farms, even though it would not permit generalization regarding the likely adoption pattern in each possible location within the country. Besides, the determination of the countrywide pattern that is likely to emerge despite regional diversity is itself a matter of considerable interest.

Accordingly, data on irrigation and farm size contained in the Sixteenth and the Seventeenth Rounds of the National Sample Survey (NSS)[8] have been pooled, averaged and presented in Table 3.1. The NSS estimates relate to the crop seasons 1959-60 and 1960-61; they have been pooled and averaged owing to the likelihood that an average of the estimates relating to two successive crop seasons may present a more reliable picture than either one taken separately. Although the estimates relate to the period 1959-61, it appears reasonable to assume that they fairly represent the basic pattern of distribution of irrigated land and of farm size on the eve of the emergence of the new technology.

Farms in Table 3.1 have been broadly divided into three categories: (I) small, (II) medium and (III) large. Although traditionally, such classification is based on the land held per farm, this measure of size has several difficulties which have been exhaustively catalogued in the voluminous literature on farm size.[9] It is recognized by and large that there is no satisfactory answer to the question of the frontiers between different farm categories, or for that matter to the questions: how small is small? and how large is large? Despite these difficulties, however, in most Indian

[8]Government of India, Cabinet Secretariat, *The National Sample Survey, Tables with Notes o.. Agricultural Holdings in Rural India*, Sixteenth Round, No. 113, New Delhi 1967; and the *National Sample Survey, Tables with Notes on Some Aspects of Land Holdings in Rural India* (States and All-India Estimates), Seventeenth Round, No. 144, New Delhi 1968.

[9]The following is a selective list of references: Raj Krishna, "The Optimum Firm and the Optimum Farm," *Economic Weekly*, October 6 and 13, 1962; A. M. Khusro, "Returns to Scale in Indian Agriculture," *Indian Journal of Agricultural Economics*, Silver Jubilee Number, 1964; V. M. Dandekar, "Foreword" in *Problems of Small Farmers*, Seminar Series, 7, Indian Society of Agricultural Economics, Bombay.

TABLE 3.1 : DISTRIBUTION OF IRRIGATED FARMS AND IRRIGATED LAND

Size-group (acres)	No. of operational farms (000)	Area operated (000 acres)	Average size (acres)	Farms with irrigation in each group (%)	No. of farms with irrigation (000)	Percentage distribution of farms with irrigation	Average irrigated land per farms with irrigation (acres)	Percentage of irrigated land in each group	Proportion of irrigated land (7/3)
	1	2	3	4	5	6	7	8	9
I.									
0.00—0.49	4838	1145	0.26	35.89	1557	6.85	0.26	0.60	1.00
0.50—0.99	4255	3076	0.72	44.72	1903	8.37	0.54	1.55	0.75
1.00—2.49	10772	18000	1.67	46.85	5047	22.20	1.24	9.43	0.74
2.50—4.99	11180	40091	3.58	49.05	5484	24.13	2.11	17.43	0.59
				(44.12)[1]		(61.55)[2]		(29.01)[2]	
II.									
5.00—7.49	6158	37015	6.01	48.02	2957	13.01	3.25	14.48	0.54
7.50—9.99	3478	29520	8.49	47.27	1644	7.23	4.25	10.53	0.50
10.00—14.99	3881	45884	11.82	46.04	1787	7.86	5.49	14.78	0.46
15.00—19.99	1843	31091	16.87	45.58	840	3.69	6.35	8.04	0.38
20.00—24.99	1111	23891	21.50	45.72	508	2.24	7.94	6.08	0.37
				(46.52)[1]		(34.02)[2]		(53.91)[2]	
III.									
25.00—29.99	663	17706	26.70	46.00	305	1.34	9.03	4.15	0.34
30.00—49.99	1121	40942	36.50	42.99	482	2.12	10.70	7.77	0.29
Above 50.00	523	38913	74.40	39.00	204	0.89	16.65	5.12	0.22
				(42.66)[1]		(4.35)[2]		(17.04)[2]	
TOTAL	49824	327277	6.56	45.61	22725	100.00	2.92		0.45

[1] Average for the group.
[2] Total for the group.

analyses, farms with less than five acres of land (regardless of quality) are taken as small farms.[10] Following this conventional usage, small farms are defined here as those with less than five acres of land, large farms as those with more than 25 acres and those with land in between these two extremes as medium sized farms. Additionally, farms in Table 3.1 have been subdivided into twelve groups on the basis of acreage.

Of immediate interest here are columns 5 and 6 showing respectively the number and percentage of irrigated farms belonging to each size-group. Essentially, these two columns show the potential pattern of distribution of farms adopting the high yielding varieties. The per cent distribution of farms with irrigation is shown in Figure 3.1; when the entire irrigated area is covered by the new varieties, the distribution of adopter farms is expected to be close to the distribution of irrigated farms as shown in this diagram.[11]

The issue posed in the literature on the green revolution—that is, whether the new varieties would be confined to a few operators of large farms to the exclusion of operators of small farms—can be answered at once. In terms of columns 5 and 6, about 14 million out of a total of 22.7 million irrigated farms, that is, 61 per cent, are small, while less than a million farms or about 4.35 per cent are large. Consequently, other things being equal, the small farms would constitute about 61 per cent of adopter farms while the large farms would constitute only about 4.35 per cent. If, as we have argued, irrigation is the main factor influencing adoption, the new varieties would not be confined to a handful of operators of large farms; instead, the new varieties would be spread over a large body of farms in which small farms would predominate.

In regard to the question as to what proportion of each category of farms could adopt the high yielding varieties, column 4 should be of considerable interest. Irrigated large farms constitute 42.6 per cent of all large farms in the country while the

[10] For instance, the Fourth Plan document defines a small holder as one with 2 hectares or less of land. See, *Fourth Five Year Plan*, p. 149.

[11] Not all irrigated land is devoted to foodgrains cultivation. However, we are disregarding this fact here since there is no information at the farm level about the proportion of irrigated land allotted to cash crops. The error involved is likely to be small, because foodgrains are grown on the bulk of the irrigated land.

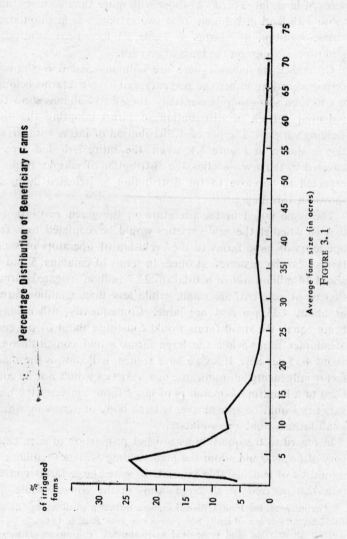

FIGURE 3.1

Percentage Distribution of Beneficiary Farms

% of irrigated farms

Average farm size (in acres)

irrigated small farms form 45 per cent of all small farms. Taking the country as a whole, therefore, the proportion of small farms that could adopt the high yielding varieties would be somewhat greater than the proportion of large farms that could adopt the new varieties.

By limiting the distribution issue to the narrow "small-vs-large farm" framework, most analysts seem to have overlooked the possibility that greater benefits could accrue to farms that are neither small nor large. As column 4 shows, 46.5 per cent of the medium sized farms are irrigated and this is the largest percentage of irrigated farms to be found in all the three broad groups of farms. The proportion of existing medium sized farms adopting the new varieties is, therefore, likely to be greater than the corresponding proportions in other groups and from this point of view, the medium sized farms as a group appear to be more favourably placed than others.

Table 3.1 sheds light on several other aspects of the issue of countrywide distribution of benefits from the new technology. For instance, the data in column 9 show that the proportion of irrigated land per farm is greatest for small farms and that this porportion is inversely related to farm size. It is, therefore, reasonable to expect that the smaller farms would have a greater proportion of land per farm under high yielding varieties thar any other farm group.

Sharing of Benefits

The expectation of a staggering increase in disparities and of polarization seems to have been based mainly on the currently inequitous distribution of land. There is no denying the fact that operated land is unequally distributed. The data in columns 1 and 2 confirm this; it is easy to see that, for instance, the bottom 18 per cent of operational farms operate slightly more than one per cent of the land, while the top 3 per cent of farms operate about 24 per cent of the area. However, in relation to the question of sharing benefits from the green revolution, it is not the distribution of operated land but the distribution of irrigated land that is relevant. The two distributions are not the same and, therefore, generalizations on the basis of the distribution of operated land would be inappropriate and could even be misleading.

To compare the two distributions—those of operated land and of irrigated land—a useful graphical device may be employed here. In Figure 3.2 the Lorenz curves for the distributions of operated land and irrigated land have been plotted together. On the horizontal axis we have taken first the cumulative proportion of operational farms; while on the vertical axis we have plotted the cumulative proportion of operated land. The diagonal straight line is the line of perfect equality; if operated land were equally distributed, the curve of distribution would lie on this line. The greater the slope of the plotted curve, or the greater the convexity of the curve toward the origin, the greater the degree of inequality. The curve of distribution of operated land is identified in the figure as curve I. Next, on the same diagram we have sketched in a curve showing the distribution of irrigated land among irrigated farms; this is identified as curve II in the diagram.

The plotted curves show that the distribution of operated land is more unequal than the distribution of irrigated land; the slope of the curve showing the distribution of operated land is greater than the slope of the curve showing distribution of irrigated land, throughout the range of observation. The latter curve lies above the former and is closer to the diagonal line of equality.

A numerical measure of the degree of inequality is the concentration ratio, which is computed as the difference between the area under the diagonal line and the area under the plotted curve. The greater the area under the curve, the closer it lies to the line of equality and consequently, the smaller is the concentration ratio. Conversely, the less the area under the curve, the further it lies from the diagonal line, and the larger is the ratio. The limiting values of the concentration ratio are one in the case of perfect inequality, and zero for perfect equality. For the distribution of irrigated land and the distribution of operated land in Figure 3.2, the concentration ratios are 0.43 and 0.59 respectively.

It is possible to go further and identify from the distribution of irrigated land the proportion of benefits accruing to different categories of farms. Column 8 in Table 3.1 indicates that about 29 per cent of the irrigated land is operated by small farms while the large farms operate about 17 per cent (see also Figure 3.3). Consequently, the share of the benefits flowing to the small farms

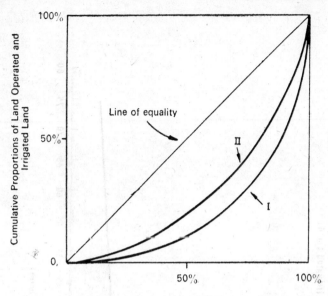

FIGURE 3.2 Distributions of Operated Land and Irrigated Land among Operational Farms and Farms with Irrigation.
I. Distribution of land among operational farms.
II. Distribution of irrigated land among irrigated farms.

should be distinctly greater than that accruing to the large farms. There is no support in these data for the view that the benefits of the new technology would be reaped only by the affluent operators of large farms while the operators of small farms would gain nothing. However, when attention is shifted to the medium sized farms, a significant fact emerges from the data in column 8. The percentage of irrigated land operated by the medium sized farms is the largest, about 54 per cent. Consequently, the share of benefits flowing to these farms is likely to be the greatest. The distribution of benefits will be skewed no doubt, but it is likely to be skewed neither in favour of the large farms nor against the small farms, but in favour of the medium sized farms.

To summarize, there is evidence that the irrigated land is not confined to a few affluent operators of large farms; instead it is

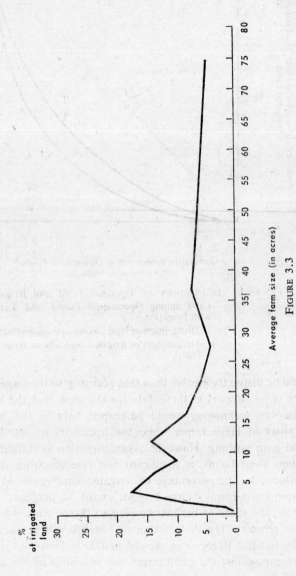

Percentage Distribution of Irrigated Land

FIGURE 3.3

widely distributed over a broad spectrum of farms; that the number of irrigated small farms is the largest; that the irrigated medium sized farms as a proportion of all medium sized farms is the greatest; that the medium sized irrigated farms operate the largest proportion of irrigated land. Correspondingly, the high yielding varieties are unlikely to be confined to a few affluent operators of large farms; they would be adopted by a broad spectrum of farms; among the adopter farms, small farms would numerically predominate; the medium sized farms are somewhat more favourably placed than other categories of farms, and although the share of benefits accruing to the small farms is likely to be considerable, the share accruing to the medium sized farms would be the largest.

Farm Size and HYVP Participation

The propositions derived here from countrywide data on farm size and irrigation cannot yet be fully tested in terms of the data provided by the evaluation studies on the HYVP; these studies in general have been location-, and/or crop-specific, designed primarily to determine what proportion of sample farms in each size-group have been participants in the HYVP. Little is known as yet about the distributions of irrigated farms and of irrigated land at the level of states, districts and villages, except that they vary considerably from the all-India pattern disclosed in Table 3.1.[12] Hence, the pattern of participation in the HYVP too is likely to vary from place to place. This is amply borne out by the evaluation studies. Thus, there is evidence that in the wheat belt, comprising Punjab, Haryana and Western Uttar Pradesh, the use of high yielding varieties has spread evenly in all size-classes of farms.[13] One early study of two blocks of Jullundur

[12] Most investigators have failed to pay attention to the distribution of irrigated land among farms. Where explicit attention was paid to irrigation, investigations have yielded data that confirm our analysis and inferences. See, for instance, C. Muttiah, "The Green Revolution—Participation by Small Versus Large Farmers," *Indian Journal of Agricultural Economics*, January-March, 1971.

[13] Government of India, Programme Evaluation Organisation, Planning Commission, *Evaluation Study of the High Yielding Varieties Programme:* (a) Report for the kharif 1967; (b) Report for the rabi 1967-68, wheat; (c) Report for the rabi 1967-68, paddy and jowar; (d) Report for kharif 1968; (e) Report for the rabi 1968-69, wheat, paddy and jowar.

district (Punjab) showed that participants were drawn from all categories of farms, and what is more important, small farms with less than 5 acres constituted about 50 per cent of the participants.[14] But in states where wheat is a minor crop, participating small farms appear to be less than the corresponding percentages of other farm groups.[15]

Again, evidence shows that in the rice growing state of Tamil Nadu, the use of high yielding paddy varieties has spread evenly across all farm size-classes,[16] although in the initial phases the small farmers appear to have been in the forefront of adoption. In the case of Thanjavur district, one early study pointed out that the "survey did not support the hypothesis that better class farmers were the only ones who benefit from such programmes of the Government."[17] Although the spread of high yielding varieties is limited in Andhra Pradesh there does not seem to be any significant relationship between farm size and participation in HYVP. Elsewhere, small farmers' participation in HYVP in respect of rice appears to have been somewhat tardy. Evidence regarding participation in the HYVP in respect of jowar, bajra and maize is somewhat sketchy and is not as extensive as that in respect of wheat or rice. There is evidence that the proportion of area under high yielding varieties per farm is the highest in the

---

Also, P. K. Mukherjee, "The HYV Programme: Variables That Matter," *E &PW* March 28, 1970; and P. K. Mukherjee and B. Lockwood, "High Yielding Varieties Programme in India—Assessment," paper submitted to Seminar on the Impact of New Techniques in Agriculture, 28th International Congress of Orientalists, Canberra, January 1971.

[14] *Planning and Implementation in Agriculture : Studies on High Yielding Varieties Programme,* vol. II, Mexican Wheat in Punjab State, Indian Institute of Management, Ahmedabad, 1967.

[15] One warning may be in order. Although the percentage of small farmers participating in HYVP in these locations may be less than that of other farm groups, it should not be concluded that the adoption of the high yielding varieties in these locations is confined to a few operators of large farms. When the proportions are translated into the number of farms in each location adopting the new varieties it will be found that numerically a greater number of small farms are using the new varieties. This is so because the farm population comprises an overwhelmingly large number of small, and a small number of large farms.

[16] *Evaluation Study, op. cit.*; also P. K. Mukherjee and B. Lockwood, *op. cit.*

[17] *Planning and Implementation in Agriculture,* vol. 1.

group of small farms and that this proportion is inversely related to farm size.[18]

By all accounts, the two states where the high yielding varieties have been widely adopted are Punjab and Tamil Nadu. Comparative data on farmer participation in the High Yielding Varieties Programme in these two states shown in Table 3.2 are suggestive

TABLE 3.2 : PERCENTAGE AREA UNDER HYV CROPS TO TOTAL
AREA UNDER THE SELECTED CROPS

State/Crop	Year and season	Below 5 acres	5 to 20 acres	20 acres and above	All sizes
Punjab: Wheat	Rabi 67-68	67.7	55.8	60.8	57.9
	Rabi 68-69	68.5	74.5	79.3	76.8
	Rabi 69-70	95.6	96·2	96.5	96.3
Tamil Nadu: Paddy	Kharif 67	65.2	55.4	77.8	60.9
	Kharif 68	67.9	51.5	31.5	48.1
	Kharif 69	57.5	54.3	68.8	58.4
	Rabi 68-69	81.8	89.1	26.6	79.1
	Rabi 69-70	74.9	78.6	67.9	75.6

Source : P. K. Mukherjee and Brian Lockwood, "High Yielding Varieties Programme in India—Assessment," paper presented at Seminar on the Impact of New Techniques in Agriculture, 28th International Congress of Orientalists, Canberra, 6-12 January, 1971.

of no bias in respect of farm size.[19] If anything, there is some indication in these data that the small farmers may indeed have been in the lead in respect of adoption of high yielding varieties at least in the initial years of the programme in Punjab; and in

[18]Michael Schluter and John W. Mellor, "New Seed Varieties and the Small Farm", *E & PW*, March 25, 1972.

[19]After his field trip in the Punjab in 1969, Ladejinsky estimated that the real sharing in the progress "is restricted to relatively few, perhaps only 10 and surely not more than 20 per cent of the farm households of Punjab." See, Wolf Ladejinsky, "The Green Revolution in Punjab: A Field Trip", *E &PW*, June 28, 1969. One should not, of course, take reports on field trips seriously. Table 3.2 shows that in Punjab, the small farmers led the adoption of high yielding varieties in 1967-68; by 1969-70, 95.6 per cent of the area operated by small farmers had been covered by the high yielding varieties. In fact, there was no significant difference among farms in respect of area under new varieties by 1969-70. Hence, Ladejinsky's contention that benefits have been restricted to not more than 10 or 20 per cent of farm households of Punjab is clearly without foundation.

Tamil Nadu, the small farmers seem to have shown a greater inclination to experiment with the new varieties and to use them more consistently over the years than other farmers. Evidence seems to indicate that, by and large, in the wheat belt, farmer participation in the High Yielding Varieties Programme has been about equal over all farms, regardless of size, but equal involvement of all farms in the programme in respect of rice has not been achieved except in Tamil Nadu and in some districts of Andhra Pradesh. This is not surprising, since the high yielding wheat varieties have been free from technical problems, whereas the new varieties of rice have been beset with technical and environmental problems right from the inception of the programme. Since the proportion of small to large farms happens to be the greatest in the rice growing areas, the participation by small farmers in the HYVP perhaps will continue to be proportionately low as long as the technical problems of high yielding rice varieties remain unresolved. Of all the high yielding rice varieties, only IR-8 and ADT-27 (an indigenous variety) seem to have fitted well into the agronomic conditions of coastal Andhra Pradesh and Tamil Nadu respectively. Widespread adoption of these two varieties in these two states suggests that in the absence of technical problems, the small farms elsewhere would not lag behind other groups in adoption.

A large body of data relating to adoption of improved and high yielding varieties exist for the districts covered by the Intensive Agricultural Development Programme (IADP). These data are valuable not only because they provide a window on the adoption pattern in these districts, but also because they indicate how a mass-action programme can involve all categories of farms. The HYVP was modelled in respect of both concept and structure, after the IADP which also involved applying a package of selected inputs in specific locations with high production potential. Like the HYVP, the IADP too is a mass-action programme, initiated first in 1960 in three districts and later extended to thirteen more districts by 1963-64. Like the HYVP again, the IADP also comprises an off-farm phase covering input supplies and an on-farm phase of demonstration of a package programme to participants. More importantly, it too is a programme that aims at involving all farmers—large, medium and small—in the districts concerned. It may be useful therefore to examine the

IADP experience in respect of farmer participation.

An early study of the IADP is the one done by Malone[20] with a representative sample of 114 farms out of some 62,700 rice farms participating in the programme in Thanjavur district of Tamil Nadu in 1962-63 during the third year of its operation. The study showed firstly, that even at that early stage of the IADP, there was widespread and equal participation by all size-groups of farms—very small, small, medium and large—and by all types of farms—operated by owners, tenants and by part-owners and part-tenants. Secondly, there was no difference among farms in respect of use of package inputs; if anything, the heavier users of improved practices were small or very small farmers. Malone concluded that the "picture is one of broad forward movement in adopting new practices in which many thousands of farmers, large to very small, jointly participate. It is not that, in military fashion, they are advancing in a straight line. But the depth of the ranks in terms of extent of use of improved crop practices per acre is not very great."[21]

A more recent study by Malone is more comprehensive in terms of coverage.[22] Analyzing the crop-cutting and other associated data for the period 1964-65 to 1968-69, from fields on nearly 5,000 IADP farms selected wholly at random from about 1.8 million farms in nine districts, he finds no evidence that small farmers have been bypassed by the process of agricultural modernization. Turning to specific districts, he finds that while in Thanjavur the proportion of small and very small farmers growing ADT-27 is greater, there is no significant difference among the proportions of different categories of farms growing high yielding varieties of maize in Ludhiana. Even in Shahabad—a district not known to be especially progressive—the proportion of very small farms growing high yielding wheat varieties is greater than the average and that of the large farms. Summarizing the evidence, Malone concludes that "there are no really important differences in the way farmers on different size farms

---

[20] Carl C. Malone, "Some Response of Rice Farmers to the Package Program in Tanjore District, India," *Journal of Farm Economics*, May 1965.

[21] *Ibid.*

[22] Carl C. Malone, "Progress in Modernization of Rice, Wheat and Maize Production in Intensive Agricultural Development Districts," The Ford Foundation, New Delhi, September, 1971.

are responding to opportunities to modernize their production methods. . . . What is documented is that the very small and small farmers, many of whom are known to be tenants are participating in the early adoption of high yielding varieties where they pay."

This broad pattern of participation in the programme is confirmed also by the study of individual districts like Thanjavur and West Godavari and Raipur made by Harrison[23] and Bouton.[24] Summing up the studies made in the IADP areas, the Expert Committee on Assessment and Evaluation concluded that the operators of small farms were in the forefront of production in many areas and were often ahead of the operators of medium and large farms in respect of adoption of new varieties and practices. While the operator of large farms have been in the lead in respect of minor foodgrains, it is the operators of medium sized and small farms who have played a more important role in respect of major foodgrains. Finally, the Committee observed significantly that "the IADP has shown an important way to involve the small farmers and thus to counteract at least partially the tendency towards unequal rate of agricultural development in modernizing agriculture.[25]

The IADP experience strongly suggests that in a government initiated mass action programme, such as the HYVP, all irrigated farms irrespective of size can be involved. It is true that success in this respect would depend at least partially on the quality of

[23] James Q. Harrison, "Small Farmer Participation in Agricultural Modernization," Staff Document, The Ford Foundation, New Delhi, December, 1970.

[24] Marshall M. Bouton, "Preliminary Report on IADP Farm and Labour Survey for Raipur and West Godavari Districts," The Ford Foundation, New Delhi 1969.

[25] Government of India, Expert Committee on Assessment and Evaluation, *Modernizing Indian Agriculture: Report on the Intensive Agricultural District Programme* (1960-68), vol. I, New Delhi 1969.

Francine Frankel's study, *India's Green Revolution (op. cit.)*, of five IADP districts uses a methodology that is questionable on many counts. Hers is a highly subjective and impressionistic account that is based on personal interviews of selected farmers and officials in a few selected villages in five districts. Unfortunately, this method of appraisal is wide open to all kinds of biases and errors entertained not only by the appraiser but also by those interviewed; and so far as impressions from field trips are concerned, it is well known that only the extremes and striking exceptions to the average and the normal are apt to catch the eye. The Expert Committee's evaluation of IADP of course flatly contradicts Frankel's appraisal.

the field organization of the HYVP. In the districts included in the IADP, a fairly adequate field organization both in terms of structure and quality had been built up over the years; these were ready to be utilized at the time of the introduction of the HYVP. Elsewhere, a satisfactory and competent organization did not exist in the initial years of the programme. This was perhaps a temporary handicap, gradually remedied as the field staff gained in experience.

Resource Constraints

We have thus far ignored the possibility that for some farms limitations on resources other than irrigation could come in the way of growing high yielding varieties, and consequently distort the adoption or participation pattern. Next to irrigation, the most important resource is operating capital with which the farmer needs to purchase seeds and fertilizer so important for raising yield. Alternatively, to procure these inputs, farmers need to have access to credit. The general impression about small farmers is that they have no surplus over consumption and therefore no saving; additionally, they are supposed to have little or no access to institutionalized agricultural credit. Credit agencies are supposed to bypass these small farmers and to confine their activities to the more affluent sections of the rural community. In view of these impressions, it is not surprising that the possibility of small farmers adopting the new varieties has been so heavily discounted in the literature on the green revolution.

Our review of the evidence on adoption has shown, however, that a large body of irrigated small farms have been using the high yielding varieties. So widespread an adoption by small farmers naturally raises the question whether the issue of resource constraints has not been blown up out of all proportion. To begin with, lack of operating capital or of credit may at the most come in the way of raising yield to the attainable level; but it is unlikely to preclude adoption of the new varieties by operators if they have irrigation. Secondly, the credit needs of participant farms in the HYVP are supposed to be worked out in advance and arrangements made for the supply of the required credit through the cooperative credit societies.[26] Under

[26]See, *Planning and Implementation in Agriculture*, Vols. I and II, *op. cit.*

the working rules of the HYVP the only farmers who could have difficulties in securing credit would be the known defaulters of the cooperatives. Admittedly, in the initial phase of the HYVP, problems were encountered in certain areas in channelling credit to participants on time, partly owing to lack of coordination among various departments and partly owing to deficiencies of the organization. But with time and experience these problems appear to have been resolved to a large extent.

There is also evidence that the small farmers' ability to finance the critical inputs of fertilizer and seeds on their own has been grossly underrated. In fact, a large number of small farmers had been using fertilizers long before the high yielding varieties appeared on the scene and opened up possibilities of large gains from fertilizer use in the cultivation of food crops. This is illustrated by the data in Table 3.3 which have been pooled and averaged from the Sixteenth and the Seventeenth Rounds of the NSS. In all likelihood, fertilizer use at that time was mostly restricted to cash crops on irrigated land. Farmers with palpably small plots of land who are supposed to grow food crops mostly or entirely for family consumption did not perhaps have the incentive to use fertilizer. Furthermore, in view of the prevailing presumption that small farmers have an extremely limited access to institutionalized credit, one would expect that the proportion of small farmers using fertilizer would be exceedingly small.

Data in Table 3.3, however, show that about 7.86 per cent of all small farms reported fertilizer use and this proportion was less than the average proportion for all farms by only two per cent. Within the group of small farms, the proportion of farms with 3.58 acres using fertilizer was above average and the proportion of farms with 2.67 acres was almost equal to the average proportion. Only in two sub-groups of small farms, those with 0.26 and 0.72 acres, was the proportion reporting use of fertilizer below average. These data seem to suggest that a large number of small farms were able to raise the necessary funds to secure whatever limited gains there could be had from fertilizer application.

The evidence in regard to fertilizer use by farms in IADP districts seems to indicate that the intensity of fertilizer use per unit of land is greatest on the small farms. However, as one would expect, there is a good deal of variation among districts in respect

TABLE 3.3 : PROPORTION OF FARMS REPORTING FERTILIZER-USE, NSS DATA

Categories	Average size of farms (acres)	Percent of farms reporting fertilizer-use
I.	0.26	3.58
	0.72	7.73
	1.67	9.81
	3.58	10.35
Average		7.86
II.	6.01	12.22
	8.49	11.70
	11.82	11.73
	16.87	12.03
	21.50	10.64
Average		11.66
III.	26.70	11.63
	36.60	10.81
	74.40	11.59
Average		11.34
TOTAL	6.56	9.88

of the proportion of gross cropped area fertilized. While in four IADP districts in the rice-growing area—Thanjavur, West Godavari, Burdwan and Sambalpur—fertilizer use is unrelated to farm size, in two districts—Palghat and Shahabad—fertilizer use seems to increase with farm size at a very low level. Again, while in wheat-growing Ludhiana there is no important difference among farms in respect of area fertilized, there is a perceptible difference in this respect in Shahabad and Aligarh. Summing up the IADP experience, Malone observes that the small farms "are active in increasing the use of fertilizer and related practices, the most available modernizing input generally in use."[27]

A more recent study[28] indicates that for the country as a whole and for most crops the proportion of gross cropped area fertilized per farm increases with farm size, but the intensity of fertilizer use per unit of cropped area is greatest on the small farms. It is possible that the proportion of the gross cropped area fertilized on small farms is determined by the area required

[27] Carl C. Malone, "Progress in Modernization" *op. cit.*
[28] "Pattern of Fertilizer Use on Selected Crops in India," NCAER, unpublished.

for growing the varieties or crops preferred for family consumption; given the small size of the holdings, the area for growing high yielding varieties for the market would necessarily be limited and this may explain the relatively smaller proportion of cropped area fertilized by small farms on the one hand, and the greater use of fertilizer by small farms per unit of land to maximize yield on the other. The same study also indicates that there is no association between availability or utilization of credit and either extensive or intensive use of fertilizer.

Evidently, a good deal of self-financing of inputs has taken place among all categories of farms and specially among small farms.[29] It appears that farmers growing high yielding varieties of rice and wheat have financed about 80 per cent of the current cost from their own funds; farmers growing new varieties of jowar have financed about 45 per cent of the costs on their own; on an average 70 to 80 per cent of the capital investment has been provided from own funds and 10 to 15 per cent have been secured as loan from private sources in paddy and wheat areas.[30]

Recent studies in some of the districts included in the IADP have revealed that some small famers have purchased land to increase the size of their holdings. One study[31] found that between 1962-63 and 1968-69, two groups of farms decreased in size and sold off and/or leased out land in West Godavari and Thanjavur districts. In West Godavari, these farms belonged to size-groups "2.6 to 5.0 acres" and "more than 20 acres"; in Thanjavur they belonged to size-groups "2.6 to 5.0 acres" and "10 to 20 acres". At the same time, farmers in West Godavari, operating land upto 2.5 acres in 1962-63 increased the size of their operational holdings by 73 per cent and their owned holdings by 69 per cent by 1968-69; the corresponding figures for Thanjavur were 16 per cent and 13 per cent. While it will not be proper to generalize from this piece of evidence alone, it certainly suggests that some small farmers, say, those with land between 2.5 and 5 acres have positive surplus or savings and unrecorded holdings

[29] John W. Mellor, "Report on Technological Advance," *op. cit.* Also, Uttar Pradesh Agricultural University, "Changing Agricultural and Rural Life in a Region of North India : A Study of Progressive Farmers in North-western Uttar Pradesh During 1967-68", vols. I and II, December 18, 1969.

[30] P. K. Mukherjee, "The HYV Programme: Variables That Matter," *E &PW*, March 28, 1970.

[31] Harrison, *op. cit.*

of cash, gold and jewellery. It also suggests that the prevailing stereotype of small farmers is in need of some modification.

This is not to suggest that small farmers have plenty of operating capital or that there is no credit problem for small farmers. We are merely arguing that the importance of capital and/or credit constraint has been exaggerated. On the whole, there is considerable support for the observation that "except possibly for intermediate term credit to finance private tubewells, credit has not yet been a substantial limiting factor to application of high yielding varieties on small holdings. . . .It is also likely that there has been underestimation of the capacity of small holder to save and invest where the returns to investment were high. Hence there has probably been a good deal of self-financing of new technologies by small cultivators."[32]

To summarize: there is evidence that while in some locations the operators of large farms may have led the adoption of high yielding varieties, small and medium sized farms have quickly followed; in others, it is the operators of small and medium sized farms who have been in the lead in respect of adoption. Adoption of high yielding varieties has not been confined to a small group of affluent operators of large farms; on the contrary, they are being used by a broad spectrum of farms in which small and medium sized farms predominate. Resource constraints do not appear to have restricted adoption by smaller farmers. Taking an all-India perspective, therefore, it is questionable that the adoption pattern has been biased against any particular group of farms.

The reasons why the emerging adoption pattern turns out to be so different from that predicted in the literature may now be summarized. First, the inherited pattern of distribution of irrigated land is not biased in favour of large farms; the small and the medium sized farms together operate the bulk of the irrigated land. Second, the average small farmer happens to be at least as progressive as any other farmer in respect of attitude toward innovation, risk-bearing and application of recommended package of inputs; contrary to the popular stereotypes, he does have positive saving however small it may be and does invest if the returns are attractive. In identifying him as backward and laggard, the critics seem to have ignored the weight of

[32] Mellor, *op. cit.*

accumulated evidence over the last two decades. Successive studies in different locations in India have pointed out that the small farmers use modern inputs more intensively on a per acre basis, and that they are eager to try out and accept an innovation if it is demonstrably superior and profitable. Finally, part of the credit at least should go to the government initiated mass-action programme under which adoption of the high yielding varieties has been taking place in India.

Adoption in the generalized model of community adoption is a free process in which farmers voluntarily and on their own initiative seek innovations to raise income; it is the end-result of farmers having moved through the stages of awareness, interest, evaluation and trial.[33] But the role of farmer initiative is limited under the HYVP; the initiative rather rests on the state governments and the field organization of the HYVP. There are targets to be filled, and specified acreage to be covered within a specified time. Past studies have indicated that the Indian farmer differs from his Western counterpart in one important respect : he innovates more under the influence of change agents than on his own rational and voluntary initiative.[34] Under a planned programme of diffusion, therefore, the pattern of adoption is likely to be different from that predicted by the generalized model. More importantly, the distinctions among early adopters, majority and late adopters, the time lag in the adoption process and the differential adoption rate that in the generalized model enables the early adopters to reap an abnormal profit, cease to be meaningful concepts. Once the decision regarding acreage to be covered in a district or block has been made, all irrigated farms irrespective of size are likely to be drawn into the programme and there would be little or no scope for an early adopter-majority-late adopter sequence.

[33] Rogers, *op. cit.*

[34] See, S. P. Bose, "Socio-cultural Factors in Farm Efficiency," in *The Indian Journal of Extension Education*, vol. I, No. 3, 1965. Also, Prodipto Roy, Frederick C. Fleigel, Joseph E. Kivlin, Lalit K. Sen, *Patterns of Agricultural Diffusion in Rural India*, National Institute of Community Development, Hyderabad (India) 1968.

Chapter Four

ASPECTS OF INTERFARM DISPARITIES

The question of income disparities being a complex one, it is perhaps desirable to consider it in several dimensions. In the literature on the green revolution, as we have noted, the question has been viewed mainly in one dimension, namely, variation in the rate and pattern of adoption of the high yielding varieties and more specifically, in terms of "differences in the level of adoption of Green Revolution technologies on small farms relative to large farms."[1] In the Indian context, differences in the rate of adoption among regions and farms may be attributed to two factors. First, the constraints on the supply of inputs, particularly in regard to certified good quality seeds, fertilizer and pesticides, restricted the area that could be covered in the initial years by the High Yielding Varieties Programme; in other words, the rate of adoption was determined initially by the supply of necessary inputs. However, in course of time, gradual improvement in the supply of inputs has rendered this constraint largely inoperative. Second, the lag in the development of adaptable varieties for particular areas has led to variations in adoption rate over a large part of the country. This constraint is the more difficult to overcome and as yet there is no sign that it will be overcome in the near future.

A variable rate of adoption, in some cases, may be a blessing in disguise. The high yielding varieties of rice, for example, represent as yet an imperfect technology. The income gains, if any, of the early adopters of the high yielding rice varieties may have been small. Correspondingly, income foregone by the late adopters may be small and insignificant. Indeed, the late adopters may

[1] William J. Staub and Melvin G. Blase, "Income Distribution and the Green Revolution," *War on Hunger*, January 1972.

realize greater profit because of fewer problems in the application of an improved innovation; the loss from failures from the use of an imperfect innovation is borne entirely by the early adopters.

Insofar as the adoption pattern is concerned, it was shown in the preceding chapter that for the country as a whole, the irrigated land is so distributed among operating farms that the country-wide pattern, despite regional diversities, would be unlikely to be biased against the small farms. And in fact, evidence does indicate that the new varieties are being used by a large body of farms, and that for the country as a whole there is no significant difference between farms of dissimilar size in respect of adoption of varieties that are free from technical problems. However, at the regional or local level, the adoption pattern would vary a good deal; and depending upon how water resources are distributed, there would be pockets of prosperity or of poverty even within a given region or state.

As a matter of fact, some increase in income disparities would appear to be inevitable not only at the local or regional level but also in the countrywide perspective. It is not reasonable to assume that only an adoption pattern biased against the small farmers would lead to an increase in disparities. As long as an innovation is restricted in its application, some increase in income disparities is unavoidable. The new agricultural technology being restricted to irrigated land, it bypasses those farmers who operate only unirrigated land. Even if it were possible to equalize the proportion of small and large farmers adopting the high yielding varieties in all locations by waving a magic wand, existing disparities would still increase, regardless of the scale-neutrality of these varieties.

It will be recalled (see chapter three, Table 3.1) that about 55 per cent of the operational farms in the country are without irrigation. These farms are not in a position to raise their income by growing the high yielding varieties. Consequently, disparities between the irrigated and the unirrigated farms are now likely to be greater than before. While both the aggregate and the average farm incomes are greater today, thanks to the green revolution, the dispersion around this new average is also likely to be greater, because income increase is confined only to 45 per cent of farms.

Thus, the answer to the question whether the green revolution is likely to increase disparities among farms is in the affirmative.

The source of this growth in inequality is the complementarity of the new technology with water. Disparities are likely to grow not because the benefits from the high yielding varieties would be cornered by the operators of large farms, but because the un-irrigated farms have been entirely bypassed by the green revolution, and because there is as yet no comparable technology to boost their income. To underscore this fact is not the same thing as to subscribe to the cliché that the green revolution would make the rich richer and the poor poorer. For, the clichè is misleading unless all irrigated farms are taken to be rich and all unirrigated farms are taken to be poor. Nearly 44 per cent of all small farms (operating less than 5 acres of land) are either wholly or partially irrigated; by no stretch of imagination can these be called rich farms. Likewise, about 58 per cent of the large farms and about 54 per cent of medium sized farms are unirrigated; but to call them poor will be to do violence to the concept of poverty itself.

Disparity Among Adopter Farms

Another aspect of the problem relates to the disparities among the subset of adopter farms. It has been alleged by a few critics that existing disparities among farms adopting the high yielding varieties have grown because the smaller farms have been unable to invest adequately in the necessary inputs; consequently, these farms have been unable to raise their yield to the level attained by the larger farms.[2] Under the circumstances, they feel, income disparities have further widened. Some have alleged that the existing differences in respect of holding size imply that the income gains, if any, for the smaller farms would be small and insignificant in comparison with the gains to the operators of larger farms.

In a sense, this is not a problem which can be blamed on the green revolution, or a problem that cannot be resolved at all. If the small farmers have been unable to invest in necessary inputs, they can be helped. A credit policy, with a pronounced bias towards the smaller operators would perhaps take care of that. Similarly, while the differences in regard to size may be difficult to reduce substantially, a fiscal policy can reduce the

[3] See, Frankel, *op. cit.*

income gains of larger operators to a tolerable and acceptable level. But the issues involved in the allegation stretch beyond the apparent apportioning of the responsibility for increasing disparities between the green revolution and what may be loosely called the agrarian structure. The issues are, first, whether the yields do differ in fact significantly among large and small adopter farms. This is a matter that can be resolved by appeal to facts, of which there are now plenty. Secondly, a related issue is that of differences in income which is far more difficult to resolve with reference to facts, because from the available data it is difficult to compute net income for different farms without taking recourse to questionable assumptions and imputations. All that it is possible to do is to derive estimates of gross income and to compare them for different sized farms. Even in regard to income, two issues may be distinguished: variations in absolute income and relative variation of income. Let us consider them one by one.

As in the case of fertilizer use so in the case of yield, there is considerable difference, crop by crop and location by location among different farms. For an overall picture, we may turn to the most comprehensive of all the available data on yield by farm size, recently presented by Malone.[3] Based on crop cutting samples from fields on nearly 5,000 farms in all the IADP districts selected wholly at random, these data relate to three crops— paddy, wheat and maize—and to five categories of farms, namely, very small (with land less than one hectare), small (with land between one and two hectares), medium (with land between two and four hectares), large (with land between four and eight hectares) and very large (with more than eight hectares of land). The over-all yields in quintals per hectare of the three crops for each farm category in 1968-69 were found to be as follows: 20.80 for very small, 20.20 for small, 21.10 for medium, 22.30 for large and 21.30 for very large farms. Expressed as per cent of the average, the yield on very small farms was 98 per cent, on small farms 96 per cent, on medium farms 101 per cent, on large farms 106 per cent and on the very large farms 101 per cent. There was, therefore, no significant difference among farms in respect of the yield obtained.

Reassuring though the evidence is, there is no gainsaying that

[3] See, Carl C. Malone, "Progress in Modernization," *op. cit.*

the absolute increase in income for small farms would be limited by the number of acres planted to the high yielding varieties, and therefore, by their small size of holdings. It will be recalled (chapter three, Table 3.1) that for about a million and half irrigated farms the average size of the total holding is no more than 0.26 acre of land; and for another 1.9 million farms with a total holding averaging 0.72 acre, the irrigated area is no greater than 0.54 acre. Whether one prefers to call them small farms or mini farms is a matter of choice, but it is unreasonable to think that the difference in the absolute increase in income between these farms and farms, say, with 16 acres of irrigated land will not be too large. In fact, the difference in respect of absolute increase in income among farm groups is bound to be striking.

While we are at it, some questions may as well be raised here, at the risk of digression, about the expectation that the green revolution would resolve the problem of low income (or poverty) of the small farms. True, the income of the irrigated small farms adopting the new varieties would improve and for some small farms with, say, 2.50 acres of irrigated land the improvement might be substantial. But the improvement would appear to be small and insufficient to resolve the problem for the majority of the operators of mini farms. To illustrate, let us assume the net income (from crop production) coefficient to be Rs.1,000 per acre of irrigated land cultivated with high yielding varieties, and further that this coefficient is double that for an acre of unirrigated land; then for the million and a half farms with 0.26 acre of land, the net income (from crop production) would be no more than Rs.260. For another 1.9 million farms with 0.72 acre of land apiece, the net income would be only about Rs.630. These incomes would be an improvement over present levels for these farms no doubt, but the improvement is hardly of a nature to make these farms affluent or to resolve their low income problem.

To return to the main issue, income disparity may be defined in several ways, but the definition that is the most important and meaningful from the economist's point of view is the one that is cast in terms of relative variation in income. Any innovation worth the name would of course tend to raise average income of adopter farms; but whether income disparities would increase would depend upon whether or not the dispersion of income around this new average also increases or not. Although the absolute income differential among adopter farms is likely to be striking now, it does not necessarily follow that the relative variation of income too would be greater.

The issues involved here may be illustrated in terms of the data in Table 4.1, compiled from a study conducted by the Agro-Economic Research Centre, Viswa Bharati, in the district of Birbhum, West Bengal in 1969.[4] In several respects, the data are incomplete and far from ideal. Information on irrigation and farmer participation is not available; data on costs are too sketchy to estimate net income. Income estimates in columns 5 and 6 are, therefore, estimates of gross income from the cultivation of paddy.

Column 5 shows realized income per participant farm from cultivation of both high yielding and local varieties of paddy. Column 6 shows what the total income would have been for the same farms if they grew only local varieties. For farms in the size-group 0.01-2.50 acres, shifting from local to high yielding varieties of paddy results in an increase in total realized income of about Rs.257 (Rs.1,310.19 less Rs.1,052.60); for farms in size-group 2.50-5.00 acres, the corresponding increase is about Rs.548 (Rs.2,680.45 less Rs.2,132.41). As one moves from the smallest to the largest size-group of farms, increase in total income owing to high yielding varieties gets larger. In the case of farms in the size-group 10.01-15.00 acres, the increase in income works out to about Rs.1,276 (Rs.8,239.64 less Rs.6,963.68). This increase in income has been expressed as index number taking the income increase in size-group 0.01-2.50 acres as base (=100), in column 7. As the indices show, the absolute increase in income varies directly with farm size. Increase in income on farms with an average of 3.91 acres of land is double that on farms with 1.93 acres while income increase on farms with 12.58 acres is about five times that on the smallest size-group of farms.

It is interesting to ask the question: what does this income increase mean to each category of farms in relation to its earlier income from cultivating local varieties? Column 8 shows that while for farms in the smallest size-group this relative increase is about 24 per cent, for farms in the largest size-group it is only

---

[4] The study has been summarized by B. K. Chowdhury in his paper, "Disparity in Income in Context of HYV," *E &PW*, September 26, 1970.

TABLE 4.1 : HYV AND INTERFARM INCOME DISTRIBUTION, BIRBHUM DISTRICT, WEST BENGAL, 1969

Size-group (acres)	Average size of participant farms (acres)	Average area per farm under HYV (acres)	Output per farm of HYV paddy (mds)	Output per farm of local paddy (mds)	Total income per farm : HYV+local paddy (Rs)	Total income per farm without HYV (Rs)	Index of absolute increase in income	Ratio of col. 5 to col. 6
	1	2	3	4	5	6	7	8
0.01— 2.50	1.93	0.55	23.12	31.20	1310.19	1052.60	100	1.24
2.50— 5.00	3.91	0.96	44.43	66.70	2680.45	2132.21	213	1.25
5.01— 7.50	5.89	1.39	52.81	69.21	2943.12	2185.03	294	1.34
7.51—10.00	9.05	1.65	78.31	169.09	5967.29	4987.77	380	1.19
10.01—15.00	12.58	1.96	97.88	243.73	8239.64	6963.68	496	1.18
Coefficient of variatio.1					133%	140%		

18 per cent. The increase as a proportion of earlier income is greatest for the farms in the size-group 5.01-7.50 acres.

What does all this mean? This means that although in absolute terms the increase in income on the smallest category of farms appears to be small, from the point of view of small farmers themselves, this increase as a proportion of earlier income is quite substantial. It means that the percentage increase in income on small farms may, in certain areas at least, even exceed that on large farms. The allegation that the green revolution does not benefit the small farms, or that it is biased toward the large farms, does not, therefore, appear to be warranted. Participant farms in Table 4.1 are presumably partially irrigated farms. As the data show, in a rice-growing district that can hardly be called progressive by any standard, partially irrigated farms with an average holding of 1.93 acres have obtained a 24 per cent increase in income by using the high yielding varieties, despite the acknowledged inadequacies of irrigation, credit facilities and supply of production inputs.   Credit for this should surely go to the new technology and to the High Yielding Varieties Programme. No other strategy, no other programme has ever offered a promise for as much to so large a proportion of small farms.

Let us finally turn to the question whether the relative variation of income has widened among the participant farms of unequal sizes. If participation in the HYVP were to increase income disparities among participant farms, then the coefficient of variation of realized income should be greater than the coefficient of variation of income derived exclusively from the cultivation of local varieties. However, the coefficient of variation of realized income turns out to be 133 per cent, whereas that for income computed under the assumption of no involvement in the HYVP works out to 140 per cent. This implies that income disparities among participant farms in this sample declined. It is true that the difference between the two coefficients is not very large, and the reduction in disparities that this difference suggests is small.   But what is important to note here is that in terms of these data there was *no increase* in income disparities among participant farms following the introduction of the high yielding varieties in Birbhum. Admittedly, in this district, the participation of small farms in the HYVP was beset with a number of problems: insufficiency of credit, unsatisfactory irrigation facilities and inadequate supply of production inputs.[5] Yet despite these difficulties faced by the small farms, it is significant that income disparities among participant farms showed a slight decline. So far as this study is concerned, the conclusion arrived at by the investigators that the HYVP further accentuated income disparity among participant farms seems to be unwarranted by the data.

The results of this study may have a greater generality because they relate to a relatively backward area and to the new rice varieties that are still technically imperfect. If in the rice growing districts despite the smaller average farm size and the unfavourable environmental conditions, relative disparity among adopter farms does not increase, then for the adopter farms taken as a whole, and for all crops, there may be no widening of relative disparity. Disregarding the coarse grains whose adoption is limited, the relative disparity is unlikely to grow in the wheat belt where the average farm size is larger, the facilities for irrigation better and the innovation itself is technically superior.

In some respects, the Birbhum study is typical of the way investigations have been conducted in India into the effects of the green revolution. The distinction between the two types of disparity has seldom been maintained and attention has been centred almost exclusively on the differences in respect of absolute increase in income among farms of dissimilar size, while wholly neglecting the question of relative variation in income. This preoccupation with absolute income differential has predictably yielded the wholly trivial result that disparities are due to the differences in the size of land holding owned or operated by different categories of farms, and this result in turn, becomes a feedback to the common belief that income disparity, or income distribution among farms and among rural households is determined by the distribution of land.[6] Because this belief so deeply influences economic investigations, analysis and policy prescriptions, it may be worthwhile to raise some issues here

[5]See, Chowdhury, *op. cit.*
[6]Cf. "Since land is a basic factor of production, the distribution of this scarce asset determines the distribution of wealth and income." Sarwar Lateef, "Land Ceilings: Questions that Came Late in the Day," *The Statesman,* New Delhi, May 22, 1972.

and to examine the alleged relationship between landownership and income distribution with the help of some data that are available.

Disparities in Respect of Land and Income

In the context of this discussion it is necessary to be explicit about the sense in which the term income is being used. For the discussion to be meaningful, it should relate to total income, rather than to one of its components, such as that part originating from crop production. Quite obviously, income from crop production is expected to be determined by land held and to vary somewhat in the same way as land is distributed. Additionally, income from crop production would vary with quality of land held and with cropping pattern. However, there is no reason why the other components of income must move in the same direction as the income from crop production, or vary with land. In the case of households without land, income from crop production is nonexistent; their income would be derived entirely from other sources. Furthermore, in the case of mini and small farms, income from sources other than crop production is likely to constitute a greater proportion of total income. If it were not so, it would be difficult to explain how operators of holdings as small as 0.24 or 0.72 acre are able to survive. In fact, these farmers have often been described as "agricultural labourers with land"—an acknowledgement of the fact that a major proportion of their income is derived from wage-paid agricultural work rather than from crop production. Traditional classification of Indian farms into small, medium and large fails to distinguish between full-time and part-time farms. Most of the mini farms and a large proportion of so-called small farms are not indeed full-time farms, but part-time farms, only partly engaged in farming; income from sources other than crop production are substantial components of their total income.

Tha this is so is amply illustrated by the farm management data.[7] So far, the study of the part-time farms has been a wholly neglected subject in India and the analysts have valiantly, if somewhat stubbornly, persisted with their analysis of farm

[7]See, for instance, Government of India, *Studies in the Economics of Farm Management*, West Bengal, Report for the year 1955-56, p. 29.

income problems entirely within the framework of the traditional classification of farms. It can be argued that the perception of the farm problems is likely to remain incomplete and may even be distorted as long as the part-time farms are not isolated from the general body of small farms; for, the problem of the two groups is not the same, though a large proportion of part-time farms do also suffer from the limitations imposed by smallness of size.

To examine the relationship between land and income distributions, we turn to the data in Table 4.2 compiled from the Rural Household Survey conducted by the National Council of Applied Economic Research (NCAER) in 1962. Agricultural households in this table refer to rural households whose major source of income is agriculture, and correspondingly, agricultural income consists of income from crop production, animal husbandry and fishery. Rural households comprise all households—agricultural and nonagricultural—and rural income includes agricultural income plus wage income from agricultural and

TABLE 4.2 : DISTRIBUTION OF INCOME AMONG RURAL AND
AGRICULTURAL HOUSEHOLDS

Income class (Rs.)	Cumulative proportion of rural households (%)	Cumulative proportion of rural income (%)	Cumulative proportion of agricultural households (%)	Cumulative proportion of agricultural income (%)
	1	2	3	4
Upto 360	6.52	1.09	4.18	0.36
361— 480	13.64	3.37	9.72	1.51
481— 600	23.31	7.34	17.48	3.57
601— 720	33.18	12.26	26.13	6.07
721— 900	46.60	20.45	39.16	11.59
901—1200	62.62	33.03	55.68	21.54
1201—1800	81.56	53.93	77.68	42.74
1801—2400	89.83	66.88	87.67	57.88
2401—4800	95.59	79.39	94.59	73.00
4801—7200	97.89	86.51	97.41	82.27
7201 & above	99.12	91.82	98.92	89.16
TOTAL :	100.00	100.00	100.00	100.00

Source: NCAER, *All India Rural Household Survey*, Vol. III, Tables 6 and 7.

nonagricultural employment, salary and income from crafts and trading.

Columns 1 and 2 show that while the share of the bottom 14 per cent of rural households in the total rural income is about 3 per cent, the share of the top two per cent is about 13 per cent. Hence rural income is unevenly distributed among rural households. Columns 3 and 4 show that the share of the bottom 17 per cent agricultural households in total agricultural income is only 4 per cent, while the share of the top 2.5 per cent agricultural households is about 18 per cent. The distribution of agricultural income too is uneven. How do they compare with each other?

To answer this question we plot in Figure 4.1 the cumulative percentages of agricultural and rural incomes against the cumulative percentages of agricultural and rural households on the horizontal axis. The Lorenz curve showing the distribution of income among rural households (identified as curve III on the diagram) lies closer to the diagonal line of equality and above the curve showing the distribution of agricultural income among agricultural households (identified as curve II) throughout its length. Therefore, the degree of income disparities among rural households is less than that among agricultural households.

To compare the distribution of income with that of land, we have also plotted in a curve (identified as curve I on the diagram) showing the distribution of land among landowning households in 1960-61. It will be noted that this curve lies the farthest from the diagonal line of equality and under curves II and III throughout its length. In comparison with the slopes of the income curves II and III, the slope of curve I is the steepest particularly around the two extreme ends. The distribution of land, therefore, is the most unequal of the three distributions. That is to say that the degree of inequality in respect of land ownership is the greatest.

The concentration ratios for the distribution of rural and agricultural incomes are 0.41 and 0.49 respectively, while the same ratio for land distribution among landowning households in 1960-61 is 0.68.[8] The presumed relationship between the

[8] The distribution of land among all rural households is predictably still more uneven. This distribution has not been sketched in on the diagram since this would have merely cluttered it up. The concentration ratio for this distribution is 0.72.

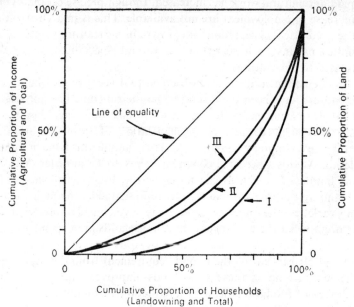

FIGURE 4.1

I   Distribution of land among landowning households.
II Distribution of agricultural income among agricultural households.
III Distribution of income among rural households.

distribution of income and of land, or between disparities in respect of income and landownership, therefore, seems to have no foundation in reality.

To conclude, the only component of total income that may vary with land, or whose distribution is likely to be governed by the distribution of land, is income from crop production. In the context of the green revolution, this component of total income is likely to vary with the distribution of irrigated land. However, what happens to total income now would depend upon what happens to the other components of total income. There is no *a priori* reason for assuming that these components of total income would remain the same. In fact, reports from areas where the green revolution has taken a firm hold indicate that employment in trade handling, transportation and marketing of grains has grown. Reports of labour shortage in the peak agricultural seasons in Punjab suggest that employment in wage-paid agricultural work has increased, though precise estimates of increase in employment are not available. This is in accord with the theoretical expectation that growth in agriculture would also induce expansion in the tertiary sector and especially in activities related to agriculture.

Some commentators have observed that the group of landless labourers have been the losers; while their income has not risen enough to compensate for increases in cost of living, they face unemployment because of mechanization of farm operations. In the next chapter we shall proceed to examine this issue in some detail. At this stage, one comment seems to be in order. While the landless households can never be the direct beneficiaries of technological progress like the one represented by the high yielding varieties, they may nevertheless receive considerable benefits indirectly. To the extent that these households are dependent on purchased foodgrains (and rural households have to buy between 30 and 50 per cent of their food requirement from the market[9]) the real income of these households improves when enhanced supplies of foodgrains lead to a reduction in foodgrains prices. Considering that a large proportion of the income of these households is spent on food alone, this benefit may be quite substantial. Obviously, it is incorrect to restrict attention to increases in money income alone and to argue that since the money income of some of these households has not been raised by a new technology, they are the losers.

[9] See, Dantwala, "From Stagnation to Growth," *op. cit.*

Chapter Five

EMPLOYMENT AND TRACTORIZATION

With about seventy per cent of the labour force engaged in agriculture, and with no sign of a decline in what appears to be an explosive rate of growth of the labour force, the problem of unemployment assumes special importance in a country like India. The advent of the high yielding varieties, however, raised expectations of an early and, one may say, a costless solution. The new varieties "generally require greater care in production and allow greater intensification, thereby providing a basis for expanded employment of agricultural labor. They can cause a shift in the demand structure towards agricultural commodities which require more labor."[1] Additionally, employment opportunities are expanded because "the possibilities of double-cropping have been made more feasible by the new genotypes, which have a shorter growing period than traditional varieties do."[2] To many observers the intensification of agricultural production with the new genotypes seemed to offer "the possibility of absorbing a considerable fraction of the growing labor force into productive employment."[3] This view is reflected in the estimate prepared by the National Commission on Labour that about half the additional labour force in Indian agriculture would be absorbed in productive employment when the targets for the HYVP and the intensive multiple cropping programme are achieved by 1973-74.[4]

[1] John W. Mellor, "Report on Technological Advance," *op. cit.*

[2] William J. Staub and Melvin G. Blase, "Genetic Technology and Agricultural Development," *Science*, July 9, 1971.

[3] Bruce F. Johnston and John Cownie, "Seed-Fertilizer Revolution," *op. cit.*

[4] Government of India, Ministry of Labour, *Report of the National Commission on Labour*, 1969.

At the same time, the bogey of a 'premature tractorization'[5] seemed to haunt the serious observers of the Indian scene, and for reasons that appeared to be quite plausible. Very substantial increases in cash income owing to the profitability of the new varieties were likely to raise the ability and the incentive of the farmers to invest in heavy labour-saving machinery and particularly in tractors.[6] Casual field trips too confirmed that in Punjab, the economy of operations and returns were uppermost in the minds of the well-to-do farmers, and that while employment had increased somewhat, a large scale displacement of farm labour particularly during the peak seasons was just around the corner.[7] Public policies too appeared to add fuel to the fire and to explicitly encourage mechanization. In 1969 the Indian government imported 15,000 tractors, 4,000 power tillers, 10,000 discs, 75 combines and a number of other items. Later it negotiated loans from the World Bank for importing 35,000 tractors and licensed additional capacity in the domestic tractor industry. Around the same time, the Planning Commission estimated the requirement of tractors at 60 to 70 thousand units per year by the end of the Fourth Plan, while the Ministry of Agriculture put it at 90 thousand units per year by 1973-74. These demand projections, sudden imports of a large number of tractors and machinery, and licensing of additional domestic capacity were factors that lent additional support to the view of a rapid mechanization of Indian agriculture.

If indeed premature tractorization did take place, the employment potential of the new agricultural technology would remain unrealized. Quite apart from that, in an economy with no structural transformation and in which the absolute size of farm labour force is rapidly growing, investment in tractors and other

---

[5] Development theory suggests that in the course of development the locus of economic activities shift from the farm to the nonfarm sector and surplus, redundant workers from agriculture are increasingly drawn into nonfarm employment. Gradually, wages rise and the price of capital declines. Finally, when the relative prices of labour and capital move in favour of capital, tractorization (or mechanization) would take place in agriculture. Tractorization, before that point in the development process is reached, has been termed "premature tractorization" in the literature.

[6] Johnston and Cownie, *op. cit.*

[7] Ladejinsky, "The Green Revolution in Punjab," *op. cit.*

machines looked difficult to justify from the society's point of view.

Employment Potential of HYV

These conflicting views on employment and mechanization have dominated the discussion on the green revolution for the last eight years. In retrospect, they do not seem to have been based on a balanced appraisal of the factors and relationships involved; they seem to lack perspective. To take up the question of employment first, it does appear that the claims about the employment potential of the green revolution were based on the assumption that the new varieties were suitable to all land and that they would cover all farms, irrigated and unirrigated. This crucial assumption, of course, had no basis in reality. The new varieties are applicable only to a fraction of the land under foodgrains; they would be confined to the irrigated areas. Consequently, increase in direct employment in agriculture would be limited. In the light of the discussion in chapter two, it should be obvious that since the quality of much of the irrigation network is indifferent, the quantum of employment generated would be yet smaller than the potential even in the irrigated areas. Furthermore, owing to the technical problems of adaptation relating to the high yielding rice varieties, the actual employment generated in the rice growing areas would be small and perhaps too small to make a dent on the problem of unemployment in the countryside. These ideas may be illustrated in terms of the state-wise data on irrigated land under cereals and on farm workers put together in Table 5.1.

Column 2 shows that some states have a very small proportion of cultivated area on which high yielding varieties can be grown. For instance, the proportion of irrigated cereals land in Maharashtra and Madhya Pradesh is only about five per cent of the total cultivated area; this proportion is about six per cent in Gujarat, eight per cent in Rajasthan and nine per cent in Mysore. Additional employment generated by the new technology in these states would be very small indeed.

Irrigated cereals acreage as a per cent of total cultivated land is the greatest in Tamil Nadu (39 per cent) and quite substantial in Punjab-Haryana (27 per cent) and Andhra Pradesh (26 per cent).

TABLE 5.1 : EMPLOYMENT POTENTIAL OF THE NEW TECHNOLOGY

States	Irrigated area under cereals (000 ha)	Col. 1 as % of cultivated area	No. of farm workers* (000)	Farm workers per 100 ha	Additional days of employment per worker per year
	1	2	3	4	5
Andhra Pradesh	3442	26	12592	98	11
Assam	571	20	2815	97	9
Bihar	2249	21	14474	133	7
Gujarat	599	6	5484	53	5
Kerala	460	17	2992	108	6
Madhya Pradesh	898	5	12188	62	3
Maharashtra	884	5	12002	63	3
Mysore	911	9	6737	65	6
Orissa	996	13	5295	71	7
Punjab/Haryana	2896	27	4168	39	29
Rajasthan	1292	8	6024	36	9
Tamil Nadu	2855	39	9060	124	13
Uttar Pradesh	4146	18	21407	94	8
West Bengal	1354	20	7249	104	8
TOTAL	23553		122487	76	8

*Farm workers include cultivators and agricultural labourers.

Source : Columns 1 and 3 from *Indian Agriculture in Brief*, Eleventh Edition.

Of these, two are rice growing states which have been relatively free from the technical problems of adaptation of high yielding varieties to local conditions; in Punjab-Haryana new varieties of wheat had no technical problems of adaptation. It is in these three regions that the additional employment generated by the new technology would be the greatest. Finally, although the proportion of irrigated cereals land is about 1/5th of the total cultivated land in the rice growing states of Assam, Bihar and West Bengal, additional employment generated in these states would be small on account of the unresolved technical problems still associated with the high yielding varieties of rice.

It is precisely in the rice growing states that the pressure of farm workers on land is the greatest. The number of farm workers per hundred hectares is 124 in Tamil Nadu, 133 in Bihar, 109 in West Bengal and 108 in Kerala (column 4). The density of farm workers is also high in Andhra Pradesh, Assam and Uttar Pradesh. These are also the states (with the exception of Assam) where the incidence of unemployment and underemployment in agriculture is high. According to the NSS, Eleventh and Twelfth Rounds,[8] the number of days of unemployment per male agricultural labourer per year owing to lack of employment opportunities was the highest in Kerala (121 days) and quite substantial in Tamil Nadu (99 days), Bihar (86 days), West Bengal (74 days), Andhra Pradesh (67 days) and Uttar Pradesh (49 days). The only rice growing state with fewer days of unemployment per male agricultural labourer was Assam (16 days).

The density of farm workers is least in the wheat growing states of Punjab, Haryana and Rajasthan. In the first two states taken together, there are 39 farm workers only per hundred hectares of land, while in Rajasthan the number is only 36. So far as Punjab-Haryana are concerned, unemployment in agriculture is lower in comparison to the rice growing states. The NSS data, referred to earlier, show that number of unemployed days per male agricultural worker per year was only 38 days in Punjab-Haryana.

What does the employment potential of the new technology mean in relation to the current work force in agriculture? To answer this question, let us assume (i) that the high yielding varieties cover the entire irrigated cereals area, (ii) that there are no technical problems regarding the use of high yielding rice varieties, and (iii) that the qualitative inadequacies of the existing irrigation network have been removed. The totality of these assumptions would impart an upward bias to the estimates of additional direct employment in agriculture, but in the context of this discussion it is preferable to err on the side of over-estimation. Furthermore, we assume that additional labour requirement of the high yielding varieties would average seventeen days per acre.[9] The estimates of additional days of employ-

[8] Government of India, Cabinet Secretariat, *The National Sample Survey, Tables with Notes on Wages, Employment, Income and Indebtedness of Agricultural Labour Households in Rural Areas*, Eleventh and Twelfth Rounds, No. 33, New Delhi, 1960.
[9] Several studies indicate that additional labour use on adopter farms

ment per worker per year that would be generated by the green revolution under these assumptions are shown in column 5 for each state separately.

It appears that increase in employment per worker would vary widely from state to state; the increase would be least in Maharashtra and Madhya Pradesh (3 days) and largest in Punjab-Haryana (29 days). In the rice growing states of Tamil Nadu and Andhra Pradesh increase in employment per worker per year is likely to be around 13 and 11 days respectively; the increase in other states would be nominal.

In terms of the current work force in agriculture, therefore, the increase in direct employment that can be generated by the new varieties is small. The proposition that the new technology has the potential to absorb a substantial fraction of the *growing* labour force in productive employment in agriculture, therefore, does not seem to be realistic at all.

Tractorization

The fear about premature tractorization was based, like the hope of an employment expansion, on the same assumption that the new technology would cover the entire area. In fact, some analysts were so concerned that the output growth would lead to a glut in the world market, precipitating ruinous price declines for foodgrains that they counselled countries like India against utilizing mechanical energy in agriculture. But since the green revolution has covered only a small area, the problem of plentiful harvest as well as the problem of labour displacement posed by tractorization is unlikely to be as serious as some commentators have made it out to be. True, high estimates of demand for tractors and sudden imports of tractors were factors that contributed to the fear of a premature tractorization. But the misgivings were based on a hasty appraisal of events. The estimates of demand for tractors soon turned out to be exaggerated and the industry became increasingly concerned about declining tractor sales; tractor imports were finally stopped altogether because of the realization that there was no gap between demand and

---

range from 10 to 25 days per acre. See, V. M. Jakhade, "Agricultural Development and Income Distribution," *Indian Journal of Agricultural Economics*, Jan-March, 1970.

domestic production. By and large, tractorization has been confined to the states of Punjab and Haryana.

Under what conditions is tractorization likely to be a widespread phenomenon in Indian agriculture? Considerations of technical efficiency may of course justify the use of tractors in some areas, as for instance, where on certain types of soil bullock power is adjudged insufficient to clear, level and plough land properly; but such tractorization of operations would be restricted to an insignificant proportion of the cultivated area and to a few farms. Considerations of prestige and status are unlikely to be of any importance as long as there is no evidence that farmers are irrational. Since economy of operations and returns is uppermost in the minds of farmers, tractorization as a widespread phenomenon can occur only in response to the operation of economic factors. Viewing the question entirely from the standpoint of a rational producer with unlimited resources, investment in tractors (and indeed any machinery) would be indicated when the value of labour saved is more than the increase in machine costs. It follows that both tractor prices and labour wages are crucial determinants of the level and the pace of mechanization. Tractor prices in India are of course amenable to public policy; and so are the wages to some extent. However, there has not been any attempt yet to raise and peg wages at an artificially high level through legislative action. In fact, increases in agricultural wages have been moderate in most of the states. The states where they have risen sharply in the last eight years or so are few, and of these states the rise has been the sharpest in Punjab, reflecting an acute shortage of labour particularly during the peak agricultural seasons. It is unlikely that relative factor prices have moved decidedly in favour of capital. In the absence of data on the relevant variables it is not possible to raise this statement from the level of assertion to the level of facts.

One variable that may reduce machine costs is clearly the size of farms; the larger the area cultivated, the less is the machine cost per unit. Consequently, it is on the fairly large sized farms, say those with more than 25 acres of land, in areas of labour shortage that tractorization is likely to take place. On these criteria, only on a very small percentage of farms can tractors be an economic proposition.

However, a minimum of tractorization may be indispensable

for output growth in certain areas. The short duration dwarf varieties have facilitated multiple cropping of land in areas where formerly it was not possible. Since the interval between the two crop seasons is short, timely completion of the required tasks becomes crucial. Here tractors and other machines can be of immense help. In fact, without these machinery, it may not be possible for the farmer to grow a second crop, or if it can be grown at all, to attain the potential yield. Faced with the problem of whether to be content with one crop or whether to grow a second crop with the help of machinery, the farmer may very well decide to tractorize operations. What appears to be a cleavage between private and social costs and/or benefits, would not seem so if indeed output has to be raised to the maximum possible limit to feed the growing population. The conflict between maximization of output and maximization of employment is there and it does not help to ignore it.

Farm Size and Labour Employment

Let us turn to the effects of tractorization on the current employment of farm labour. Apprehensions about increasing unemployment of farm labour arise generally from a widely held belief that large farms alone provide wage-employment and that other farms are self-employed farms. The Second Agricultural Labour Enquiry, for instance, observed that "no distinct employer-employee relationship exists in agriculture except in the case of big landholders and the farm hands they employ."[10] And again, "Two factors severely limit the scope for wage-employment of agricultural labourers. They are: use of family labour on the farm and mutual help prevalent in agriculture. The small farmer economizes, to the extent he can, by making members of his household work on the farm, and it is only when they, as a team, are unable to cope with rush work that he employs hired labour."[11]

This view, though widely held, is inconsistent with the accumulated evidence of over two decades. Farm level data show that workers are hired both on long-term (permanent) and short-term (casual) basis on all size-group of farms in all states. In

[10] Government of India, *Second Agricultural Labour Enquiry*, p. 37.
[11] *Ibid.*, p. 39.

Table 5.2, data from NSS, Seventeenth Round are presented on permanent hired workers. It will be noted that about 2.3 million small farms, with holdings of less than five acres, employ altogether 3.8 million (or 28 per cent) of workers on a peramanent basis; about 52.6 per cent of all permanent workers are employed by medium sized farms with land between 5 and 25 acres. These two categories of farms taken together employ about 81 per cent of all permanent workers. In constrast, farms with more than 25 acres employ only 19 per cent. These data would seem to call for a revision of the conventional view about the relative importance of different categories of farms as employers of hired labour.

So far as casual labour is concerned, comparable data are not available; but the basic facts can be gleaned from the data on hired casual labour days both on per farm and on per acre basis provided by the farm management surveys. As a typical example, average casual hired labour days used per farm in two

TABLE 5.2 : SIZE OF FARMS AND PERMANENT HIRED WORKERS

Farm size (acres)	No. of farms with attached workers (000)	Average No. of attached workers per farm	Total No. of attached workers (000)	Proportion of attached workers (%)
	1	2	3	4
I				
0.01— 0.49	85	1.51	128	0.9
0.50— 0.99	201	1.46	293	2.2
1.00— 2.49	714	1.69	1207	8.9
2.50— 4.99	1299	1.71	2221	16.4
II				
5.00— 7.49	1058	1.63	1724	12.7
7.50— 9.99	677	1.76	1191	8.8
10.00—12.49	596	23.6	1407	10.4
12.50—14.99	343	2.02	693	5.1
15.00—19.99	506	2.29	1159	8.5
20.00—24.99	348	2.42	842	6.2
III				
25.00—29.99	227	3.13	710	5.2
30.00—49.99	411	2.46	1011	7.4
50.00 & above	258	3.37	869	6.4

districts of Uttar Pradesh in 1956-57 are shown in Table 5.3. It will be noticed (in column 1) that the number of casual labour days used per farm increases with farm size; the number of days is the least (only 15.3 days) in the case of farms below 2.5 acres and as high as 236.1 in the case of the large farms with more than 25 acres in farm. If the data are taken as typical for the state concerned, the relative importance of each size-group of farms as employers of casual workers can be estimated for the state as a whole. As an illustration, such an estimate has been developed for the basic data in columns 1 and 2. The number of operational farms in each size-class (column 2) is estimated from the NSS data, Seventeenth Round. In all, close to 520 million casual labour days may have been hired by the farms, but as column 4 shows, about 42 per cent of 520 million casual labour days may have been hired by the small farms with land less than five acres per farm, and about 51 per cent by the medium sized farms, while the proportion hired by the large farms (with land

TABLE 5.3 : UTILIZATION OF HIRED CASUAL WORKERS IN UTTAR PRADESH

Farm size (acres)	Hired casual labour days per farm	Operational farms (000)	Total casual labour days hired (in million days)	Proportion of casual labour days hired by farms in each size-class
	1	2	3	4
0.01— 2.50	15.3	4583.9	70.1	13.5
2.50— 5.00	52.6	2807.7	147.7	28.4
5.00— 7.50	37.9	1379.5	52.3	10.0
7.50—10.00	77.8	647.4	50.4	9.7
10.00—15.00	112.4	647.4	72.8	14.0
15.00—20.00	252.4	241.2	60.9	11.7
20.00—25.00	255.7	112.1	28.7	5.5
25.00 & above	236.1	157.4	37.2	7.1
		10579.0	520.1	100.0

Source : Column 1 is derived from data in *Studies in Economics of Farm Management in Uttar Pradesh*, Report for 1956-57, Government of India, Ministry of Food and Agriculture, Directorate of Economics and Statistics, New Delhi, p. 16.

Column 2 is derived from *Tables with Notes on Some Aspects of Landholdings in Rural Areas* (State and All-India Estimates), Seventeenth Round, September, 1961—July 1962, The National Sample Survey, No. 144, 1968.

more than 25 acres) may have been no more than seven per cent. As in the case of permanent workers, therefore, the medium and the small farms seem to employ a greater proportion of casual workers, and their relative importance as employers of hired workers is far greater than the large operational farms.

The basic pattern of hired labour utilization revealed by the data from Uttar Pradesh has a widespread generality. Examination of farm management data in different regions would yield essentially the same pattern. Harrison's study in West Godavari and Thanjavur districts too confirm the proposition that the relative importance of farms in the top size-classes as employers of hired labour is small; according to this study, more than 80 per cent of hired labour is employed on farms with less than 20 acres in West Godavari and Thanjavur districts.[12]

Given this pattern of hired labour utilization, it seems reasonable to conclude that only a small proportion of hired labour may be affected if farms in the top size-classes were to turn to tractors; but at the same time, if the shift to the new technology raises labour requirement in all size groups of farms then labour displaced on large farms may well be absorbed on farms in other size-classes.

To summarize, although on the one hand, large farms employ greater number of workers on a per farm basis, the number of large farms is small; out of a total of 49 million farms in 1960-61, farms with more than 25 acres amounted to only about 2 million. Hence, large farms altogether employ a small proportion of farm labour. On the other hand, employment of hired labour on farms below 25 acres is small relative to that on large farms; but the number of farms with less than 25 acres of land being large, a relatively large proportion, in fact, the bulk of the farm labour is employed on these farms. Hence, the employment prospects for farm workers in the context of the green revolution would depend not so much on the labour-use decisions of large farm operators, but on the conditions of demand for farm labour on small and medium sized farms.

[12] Harrison, *op. cit.*

Chapter Six

LAND REFORM
Anatomy of a Myth

The search for the means to resolve the problems of equity and welfare in the context of the green revolution leads one inevitably to the consideration of land reform. This is not surprising, for, the measure is popularly believed to be a panacea for all equity and welfare problems in agriculture and additionally, supposed to be relatively simple and cost free.[1] Most of the arguments for land reform in the present context, as we shall see, are either old or mere extensions of those which have been repeatedly used over the last two decades or so in other contexts. Therefore, an evaluation of the measure now will be impossible without reopening old issues or without taking a fresh look at the premises on which the arguments rest. But before we proceed to sift the chaff from the grain, we need to clear up a semantic problem.

The problem relates to the connotation of the term 'land reform', which seems to mean different things to different people. This is not surprising, for, the way the term has been defined in the literature is much too broad to be precise. According to Raup,[2] it may mean, on the one hand, any of the measures directly involving the tenure under which land is held : promotion of ownership by operator and the reduction of absentee landlordism; regulation of rents and the legislative protection of security of tenure; consolidation of fragmented holdings;

[1] Unfortunately, the cost of land reform has seldom been a worthwhile subject of scrutiny for academic pundits. Important though the subject is, we are unable to consider it here for lack of space.

[2] See, Philip M. Raup, "The Contribution of Land Reforms to Agricultural Development : An Analytical Framework," *Economic Development and Cultural Change*, October, 1963.

subdivision of large holdings; control of land inheritance to prevent subdivision or to discourage land concentration; improvement of land surveys and land records. It may also mean, on the other hand, any of the measures not directly related to land tenure, such as: development of an agricultural extension service; reform in land tax and fiscal policies; improvement in the marketing system, conditions of work of agricultural labour, agricultural credit system and land market.

Such a broad definition is a veritable source of confusion. In the discussion that follows, the term 'land reform' will be used specifically to refer to a single measure, namely, imposition of a ceiling on land holdings and subsequent redistribution of land to the landless in the countryside. Such a restricted use of the term will, it is hoped, make for clarity and, in addition, will make explicit the identification between land reform and land distribution to the landless that has come to be firmly established in common parlance at least in India. With this matter of definition straightened out, let us now turn to the arguments stressing the need for land reform.

Land Reform and Incentive for Growth

To begin with, there is the question of an appropriate economic climate conducive to the spread of the high yielding varieties. Brown[3] has argued that the reason for the varying impact of the 'seeds of change' in different countries is "the inability of some governments to create the economic climate required to accommodate a breakthrough in agriculture—more specifically, to their inability to bring about land reform. Even if price incentives are provided, land ownership in many countries is too concentrated within a small segment of the population to permit an effective link between effort and reward for those who work the land." In order to create the climate for a wider diffusion of the new technology, he has urged the governments of these countries to redistribute land and to broaden the base of landownership.

This argument entirely overlooks the complementarity between the new technology and irrigation and assumes, incorrectly, that the new varieties have universal application.

[3] Brown, *op. cit.*, pp. 110-111.

The factor responsible for differential spread of the high yielding varieties among countries and among regions within countries is the disparate development of water resources rather than tenurial arrangements. The most impressive spread of the new varieties in India has occurred, as Dantwala points out,[4] in Punjab, Haryana and Tamil Nadu, that is, in states not particularly distinguished either in respect of land reform legislation or in respect of its implementation. Obviously, land tenure arrangements in these states have not come in the way of a widespread adoption of the new technology. In contrast, the progress of the green revolution has been tardy in Maharashtra and Gujarat—states which have a better record in respect of both legislation and implementation of land reform.

Brown's argument is in fact an offshoot of the more general and familiar proposition that land reform is essential for agricultural growth. Under certain circumstances of course, land reform could conceivably lead to an optimum land utilization and to an expansion of output. As Doreen Warriner has pointed out, if the resource base of agriculture is adequate, if reserves of productivity exist, and if the major constraint is only institutional tenure arrangements, as is the case in Latin America, land reform could be expected to initiate agricultural growth.[5] Farm output could also grow, if appropriate technological innovations were available. None of these conditions is fulfilled in the Indian case. Reserves of productivity do not exist that can be exploited with ease. Furthermore, recent technological innovations offer little promise for raising yield on the unirrigated land that comprise about three-fourths of the cultivated area. It is unreasonable to expect under the circumstances, that the beneficiaries of land reform would be able to raise yields above present levels.[6]

[4] See, M. L. Dantwala, "From Stagnation to Growth : Relative Roles of Technology, Economic Policy and Agrarian Institutions," Presidential Address at Fifty-Third Annual Conference of Indian Economic Association, Gauhati, December 1970.

[5] Doreen Warriner, *Land Reform in Practice and Principle*, Oxford, 1969.

[6] The problem of Indian agriculture has been one of inadequacy of the resource base and absence of reserves of productivity. The green revolution, limited though it is in extent, has forcefully underscored the fact that the major constraints on Indian agricultural growth have been technical rather than institutional. See, Dantwala, *op.cit.* Also, W. David Hopper, "The Mainsprings

Let us cast a closer look at the hypothesized relationship between land reform and agricultural growth. The proposition that land reform would inevitably lead to an increase in farm productivity and output derives support in no small measure from the miraculous powers attributed to the psychological factor of ownership which, according to the old adage, 'turns sand into gold'.[7] One cannot, however, set much store by this essentially European view of ownership and enterprise. Indeed a close look at the Indian farms would strongly suggest that simple generalizations and too uncritical extrapolation of the general European experience would be misleading. The relationship between ownership and enterprise, or between ownership and investment is not simple here; owned farms are often neglected while tenant farms are often marked by superior performance. In fact, the psychological factor of ownership has not helped solve the low income problem of million upon million of owner-operated Indian small farms. An additional question here is whether the beneficiaries of land redistribution who have had no experience in farming and management can be expected to turn out as entrepreneurs just because of the psychological incentive provided by ownership. As Doreen Warriner pointed out long ago, "the peasant economies of Western Europe have evolved in very favourable natural conditions and in a very special historical environment. These conditions have enabled them to reach their present high level of productivity through gradual improvement in grain yields and cattle raising. But the peasant is not a basic universal type, a sociological constant. Where cultivators have been small tenants, and have had no example of better farming to follow, we cannot expect that

---

of Agricultural Growth," Dr. Rajendra Prasad Memorial Lecture to the 18th Annual Conference of the Indian Society of Agricultural Statistics, January 28-30, 1965.

[7] Wolf Ladejinsky, "Land Ceiling and Land Reform," *E&PW*, Annual Number, February 1972. Ladejinsky also cites the Indian farm management data of the mid-fifties in support of the related proposition that small farms are more productive than the large farms. This is hardly the place to reopen the debate on interfarm comparison of productivity, but it needs to be pointed out that the evidence provided by the data in the fifties is far from conclusive. As Rudra has shown, there is reason to believe that the inverse correlation between yield and farm size in these data are spurious. See, Ashok Rudra, "Farm Size and Yield Per Acre," *E&PW*, Special Number, July 1968.

ownership will suddenly transform them into real farmers."[8] The same argument applies with perhaps greater force to the case of recipients of small plots who were formerly landless.

The Japanese and the Taiwanese experiences are often cited in support of the proposition that land reform would lead to agricultural growth. There are, however, conceptual difficulties in attributing the spurt in agricultural productivity and output witnessed in these two countries wholly or even partially to land reform. It cannot be denied that there is an apparent association between land reform, increased technical inputs and increases in fixed and working capital on the one hand and productivity and output expansion on the other in these countries because these happened to occur at the same time; but it is questionable if the association between land reform and increase in productivity is real. The kind of increases in fixed and working capital that took place in these two countries concurrent with land reform, would have led to output and productivity expansion of the same kind anywhere without land reforms; witness for instance, the experience in Indian Punjab during the period 1950 to 1965. Although ceilings legislation in the fifties failed to generate a significant amount of surplus land in the Punjab for redistribution, the growth of Punjab's agricultural output has been highly impressive. From a base of 100 in 1950-51 the index of agricultural production rose to 210 in 1964-65. In the same period there was almost a four-fold increase in the use of newer inputs of production such as fertilizer, pesticides, electricity, fuel oil and water. A cause and effect relationship between land reform and productivity increases would be difficult to establish only on the basis of historical experience in a couple of countries.

In an important sense, the example of land reform in Japan and Taiwan is not quite relevant to this discussion. The nature and the substantive content of land reform in these two countries were different. Its objective was "land to the tiller" rather than "land to the landless". Land reform, therefore, entailed a change in property relations—a transfer of ownership rights from non-cultivating owners of land to the actual cultivators who had

[8] Doreen Warriner, "Land Reforms and Economic Development," in Eicher and Witt (eds) *Agriculture in Economic Development*, McGraw-Hill, 1964.

already been in possession of the land as tenants; it also involved, along with a transfer of ownership title, an immediate transfer of rental income from landowners to the erstwhile tenant cultivators. But it did not involve splitting up of the production base of the farms; the size of units of operation and management remained the same.[9] The worker/land ratios on already cultivated land remained undisturbed, while additional workers were absorbed in farming only through the sale of public land.

Since the type of land reform we are concerned with invariably involves splitting up of larger holdings and absorption of additional labour on already cultivated land, there is a likelihood of decline in output per worker,[10] even if output per acre were to remain unchanged. And indeed the prospect of a progressive decline in output per worker over time would appear to be very strong if viewed in the dynamic context of unabated growth in the number of workers in agriculture. Any suggestion that increase in output per worker should be an important consideration in formulating agricultural policies is apt to invite the derisive comment that the concept is borrowed from the Western experience, or that the concept is a cover for enlarging farm size through collectivization.[11] This is unfortunate, for the concept of increasing output per worker has justification quite independent of the merits of mechanization or of enlargement of farm size. The goal of development being increase in output per capita, one cannot ignore the likely effects of recommended policies on output per worker. Maximization of output per unit of land may be an appropriate goal in a stagnant economy with no desire or prospect of growth whatsoever. Output per unit of worker in agriculture must grow in an economy that seeks development and improvvment in welfare and income per worker; and it can be achieved only by ensuring a steady trickle of labour outflow from agriculture.

[9] See, Takekazu Ogura (ed) *Agricultural Development in Modern Japan*, Fuji Publishing Co. Ltd., Tokyo, 1963; and Raymond P. Christensen, *Taiwan's Agricultural Development: Its Relevance for Developing Countries Today*, U.S.D.A., E.R.S.; 1968.

[10] R. S. Eckaus, "Technological Change in the Less Developed Areas" in *Development of the Emerging Countries : An Agenda for Research*, The Brookings Institution, Washington D.C., 1962.

[11] See, for instance, Edgar Owens and Robert Shaw, *Development Reconsidered*, D.C. Heath & Co., Lexington, 1972, p. 59.

This is not to deny the importance of increasing output per acre; clearly yields need to increase. But increase in yield is not the only criterion for evaluating land reform. From the broader point of view of the economy as a whole, there are other considerations. One of these is the likely effect on output per worker. Another important consideration is whether saving and marketable surplus of foodgrains would decline as a result of redistribution of land to the landless.[12] If an increased proportion of total output is consumed as a result of its distribution over a larger number of workers, total marketable surplus and total saving may be reduced. Such a development would impede further progress. Unfortunately, in their reformist zeal, most writers have totally ignored these considerations of crucial importance to the development seeking economies.

Land Reform and Equity

Land reform has been urged also as a means to ensure an equitable sharing of the benefits of the new technology. As Shaw puts it, larger farmers "have received the benefits of the Green Revolution at almost no cost to themselves. . . .It is only reasonable then that the larger farmers should share some of the benefits with their poorer neighbours. Land reform is one way to achieve this."[13] This view, like the one we have already examined, takes the universality of the green revolution for granted. As we have noted, the green revolution concerns only the irrigated land; if the benefits of the new technology are to be spread evenly among all rural households—farmers and landless workers alike—through a redistribution of land, it is the irrigated land that needs to be redistributed. But since irrigated land is so unevenly distributed over space, in most areas there would be nothing to distribute.

Essentially, Shaw has merely extended the case for redistribution of land in general to secure social and distributive justice

---

[12] The possibility of a decline in marketable surplus of foodgrains following land redistribution emerges forcefully from Raj Krishna's study, "The Marketable Surplus Function for a Subsistence Crop : An Analysis with Indian data," *Economic Weekly*, February 1965.

[13] Robert d'A Shaw, *Jobs and Agricultural Development*, Overseas Development Council, 1970, pp. 60-62.

or to reduce disparities. As Ladejinsky puts it, "At the bottom of the ceiling advocacy are the wellknown inequalities of the Indian agricultural structure and the increasing dependence upon land for livelihood by more than 70 per cent of the population. . . .The more telling argument is the fact that in Indian conditions with all too narrow an industrial base, ownership of land, or for a tenant to remain secure on the land, is the minimum security an underprivileged farmer can look forward to. This is the principal economic justification of the ceiling."[14]

One has to concede that a ceiling programme, properly implemented, could reduce inequalities in the countryside. Since land is unevenly distributed and since more than eight million households (in 1960-61) are landless, a ceiling and redistribution programme, regardless of the ceiling level, would indisputably reduce inequalities in respect of ownership of land. This reduction, of course can only be nominal, given the availability of land. To illustrate with 1960-61 data, if 6 million acres were acquired from owners with more than 50 acres of land and allocated to 8.5 million landless rural households, the concentration ratio, measuring the degree of inequality in distribution of land among rural households would decline from 0.72 to 0.69—a decline of about 4 per cent.[15] With less amount of surplus land to redistribute, the reduction in inequalities, or what comes to the same thing, the improvement in land distribution will be smaller. However welcome such a small reduction in disparities may appear to be, it is obvious that it would also proliferate the number of low-income poverty-stricken farm households instead of reducing their number. It is also possible that the reduction in inequality may turn out to be more illusory than real, if past experience is any guide. Parthasarathy and Raju have pointed out that redistribution of very small plots of land to the landless, "however politically attractive is least desirable.

[14] Ladejinsky, *op. cit.*

[15] To determine the equity effects of land reform programme it is necessary to make use of regional data. Since our purpose here is to suggest the magnitude of decline only, we have used the aggregate data. Our illustration is also impractical. The ceilings legislation passed in 1972 by any standard is rather constrictive. And yet, it is estimated to yield on an optimistic reckoning no more than 3.3 million acres of land as surplus. Whether it will be possible to generate 6 million acres or more as surplus land through a more constrictive programme is a question that need not hold us here.

Past experience of assignment of government land to the landless has shown in many places that such land is either leased-out or, much worse, sold for a song to others."[16]

The process of land transfer may not be instantaneous; there may therefore be some improvement in land distribution in the short run. Whether this improvement will be enduring or can be maintained over time is another matter. Essentially, a ceiling and land redistribution programme is a static solution of the equity problem in a static setting. For a *constant* population in agriculture, or what amounts to the same thing, for a *constant* number of rural households it is conceivable that the degree of equity resulting from land redistribution in the present could be maintained over time. But viewed in the dynamic context of population growth, subdivision and proliferation of households, efforts spent on land redistribution appear to be vain and self-defeating exercises in equity. This is certainly an important issue that has been totally ignored in the literature on land reform.

Tenancy: Facts and Fictions

Several observers have held that "the high yielding varieties, particularly when combined with high output prices, encourage eviction of tenants and resumption of cultivation by landlords. There are a number of areas in which future success of a high yielding varieties program may shift a substantial number of people from the status of tenants to the status of landless labourer."[17] Others have predicted that this process would lead to the emergence of a bimodal distribution of operational farm size.[18] To prevent tenant displacement, premature mechanization and consequent income disparities, therefore, "a broader base of land ownership" has been recommended by these writers.[19]

This is not a very convincing view, however. For one thing, there is no evidence as yet that the 'success' of the High Yielding

[16] G. Parthasarathy and K. S. Raju, "Is There an Alternative to Radical Land Ceiling?" *E &PW*, July 1, 1972.

[17] John W. Mellor, "Report on Technological Advance," *op. cit.* See also Robert d'A Shaw, *Jobs and Agricultural Development, op. cit.*

[18] Bruce F. Johnston and John Cownie, "The Seed-Fertilizer Revolution," *op. cit.*

[19] John W. Mellor, *op. cit.*; Shaw, *op. cit.*

Varieties Programme has led to the eviction of tenants on a large-scale, or its corollary, to the resumption of land by owners of land on a large-scale; and for another, a land redistribution programme is extremely unlikely to prevent mechanization, since a farmer can always lease-in land to raise the size of operated holding sufficiently to reap the benefits of machinery. Besides, if large-scale mechanization and eviction of tenants were to occur as a consequence of the green revolution, the corrective measure most appropriate would appear to be not land redistribution but what Dantwala so aptly calls 'protective legislation' to provide protection to tenants and sharecroppers.[20]

Basically, concern about tenant displacement appears to spring from a misreading of the magnitude and the extent of tenancy in India; and it seems to draw considerable support from the conventional view of tenancy. According to the conventional view, there is a land surplus sector in which a few landlords own large plots of land and a land-scarce sector in which too many farmers own small plots; the pressure of population is unevenly distributed over the two sectors. Landlords in the land surplus sector, interested in the maximum exploitation of tenants, lease-out their holdings in small parcels to tenants in the land-scarce sector, while the tenants in their turn subsist by exploiting the land.[21] Given this conventional view, fears about large-scale eviction of tenants and the emergence of bimodality in the farm size distribution may not be wholly unjustified in the context of the green revolution.

In another context, Doreen Warriner has observed that the study of land reforms is impeded "because we know too much and think too little."[22] That we think too little would be readily conceded; but it needs to be questioned whether indeed we know too much. In retrospect, the ever-growing literature on Indian tenure system and land reform seems to have been based on a very shallow foundation of facts. What masquerade as facts are often no more than untested hypotheses or propositions. To what extent is the conventional view of tenancy justified?

---

[20] M. L. Dantwala, "From Stagnation to Growth," *op. cit.*

[21] This view finds its finest expression in the paper by V. M. Dandekar, "Agricultural Growth with Social Justice in Over-populated Countries," *E&PW*, Special Number, July 1970.

[22] Doreen Warriner, "Land Tenure in Kerala," *E&PW*, July 10, 1971.

The Seventeenth Round data of the NSS show that the extent of tenancy is limited.[23] Leased-in area as per cent of total operated land was only 10.70 in 1960-61, while the farms leasing-in land formed about 23.52 per cent (18.38 per cent part tenant-part owner farms plus 5.14 per cent 'pure tenant farms') of the operational farms. Of the farms which leased-in some land, about 36 per cent were medium sized and large farms each operating more than five acres of land; furthermore, these 36 per cent farms operated more than 73 per cent of the leased-in area. In contrast, small farms with less than five acres of land leased-in only about 27 per cent of the total area under tenancy. Obviously, tenants are not homogeneous as a group; nor are all tenant farms small farms. The generalization that tenants come from the 'land scarce sector' appears at best to be a partial truth; the number and the proportion of tenants from the 'land surplus sector' are quite considerable. Although the importance of tenancy or leasing-in of land by small farms as a means of raising income levels is undeniable, its importance cannot be inconsequential for the medium sized and large farms which together operate close to three/fourths of the total leased-in area.

The view that landlords in the 'land surplus sector' lease-out land in small parcels to tenant operators in the 'land scarce sector' implies that the distribution of operational farms and the distribution of landowing households are not alike; in particular, it implies that the number and the proportion of operational farms are (i) considerably greater than the number and the proportion of landowning households in the smaller size-classes and (ii) considerably smaller than the number and the proportion of landowning households in the larger size-classes. This implication, however, is not borne out by the data that are available. A comparison of Figure 6.1 depicting the distribution of landowning households with Figure 6.2 showing the distribution of operational farms would suggest that the two distributions are remarkably alike; however, the Figures do not fully bring out the fact that in the smaller size-classes the number

---

[23] Government of India, Cabinet Secretariat, *The National Sample Survey, Tables with Notes on Some Aspects of Land Holdings in Rural Areas* (States and all-India Estimates), Seventeenth Round, September 1961-July 1962, No. 144, New Delhi 1968. See also, Dharm Narain and P.C. Joshi, "Magnitude of Agricultural Tenancy," *E &PW*, September 27, 1969.

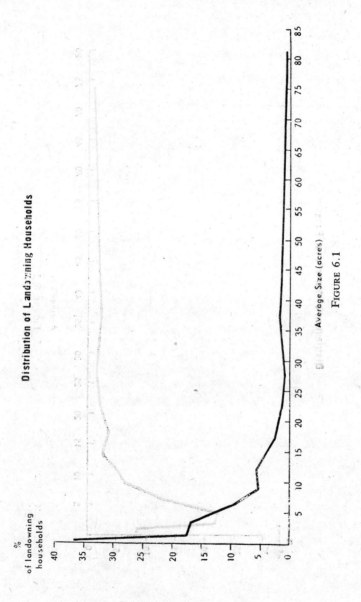

Distribution of Landowning Households

%
of landowning
households

Average Size (acres)

FIGURE 6.1

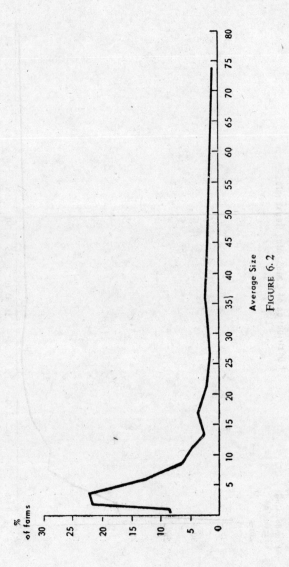

Distribution of Operational Farms

Average Size

FIGURE 6. 2

and the proportion of operational farms are smaller than the number and the proportion of landowning households, and that an opposite pattern obtains in the larger size-classes. In the Seventeenth Round, NSS data, on which these diagrams are based, there were 31.3 million or 61.69 per cent operational farms as against 46.0 million or 75.22 per cent landowning households in the small size category (that is, each operating less than five acres of land); in contrast, there were 2.3 million or 4.52 per cent operational farms against 2.1 million or 2.85 per cent landowning households in the category of large size (that is, each operating more than 25 acres of land). Significantly, while the area operated by small operational farms was in close correspondence with the area owned by small landowning households, the area operated by large farms exceeded the area owned by large households by about 5.6 million acres.

These data suggest that leasing-in of land is not confined to the small operators alone; nor is the leasing-out a practice indulged in only by the large landowners. All categories of farms—small, medium and large—lease-in and lease-out land. What is more, the importance of leasing-out of land as a source of income may be greater for the small landowners. To show this clearly, one needs to turn to the Eighth Round, NSS data, which provide information on leasing-out of land by size-classes of households. In Table 6.1, compiled from the above-mentioned source, the percentages of households leasing-out land vary from 8.13 in the smallest size-class to 36.26 in the largest size-class (column 2). The actual number of households leasing-out land in each size-class and the percentage distribution of these households are given in columns 3 and 4 respectively. These two columns provide an answer to the question: what proportion of the households leasing-out land belongs to the category of large, medium and small landowning households? It will be noted (column 4) that the percentage of the number of households leasing-out land rises sharply as one moves from the smallest size-class (less than 0.49 acres), reaches the peak in the size-class 1-2.49 acres and then declines. In terms of the broader definition followed in this book, of the households leasing-out land about 54 per cent are small landowning households, about 37 per cent are households with medium sized holdings and about 9 per cent are households with holdings larger than 25

TABLE 6.1 : HOUSEHOLDS LEASING-OUT LAND, ALL INDIA

Size of household ownership holding (acres)	Total number of households (000)	Percentages of households leasing-out land	Number of households leasing-out land (000)	Percentage of number of households leasing-out land
	1	2	3	4
I 0.01— 0.99	15360	8.13	1249	16.34
1.00— 2.49	8879	16.31	1448	18.96
2.50— 4.99	8569	16.64	1426	18.66
			(4123)[1]	(53.96)[2]
II 5.00— 7.49	4966	18.43	915	11.98
7.50— 9.99	2972	18.54	551	7.21
10.00—14.99	3207	22.89	734	9.61
15.00—19.99	1690	23.85	403	5.27
20.00—24.99	929	25.94	241	3.15
			(2844)[1]	(37.22)[2]
III 25.00—29.99	636	25.00	159	2.08
30.00—49.99	1051	28.07	295	3.86
50.00 & above	604	36.26	219	2.88
			(673)[1]	(8.82)[2]
TOTAL	48863	15.64	7642*	100.00

* Figures do not add up owing to rounding.
[1] Total for the group.
[2] Total for the group.
Source : *The National Sample Survey*, Eighth Round : July 1954-April 1955, No. 36, Report on Land Holdings (3) (Some Aspects of Ownership Holdings).
(1) Columns 1 and 2 from Table 6.7, p. 30.
(2) Column 3 from Table 29.
(3) Column 4 from Table 36.

acres. It appears, therefore, that the bulk of the rent-receivers are the small landholders for whom the importance of leasing-out may be greater than that for any other category of land-owning households. There is no evidence in these data of the relation between holding size and leasing-out postulated in the conventional view of tenancy.

Land Reform and Labour Absorption

We turn now to examine the case for land reforms as a means to absorb more labour in agriculture. Several analysts have argued that since small farms are more labour intensive than the large farms, a ceiling on landholdings and redistribution of land in small plots would resolve the problem of unemployment in rural areas arising out of mechanization and indeed would encourage a pattern of agricultural development which preserves a unimodal distribution of farm sizes. Such a strategy, tested and found effective in Japan and Taiwan, would be justified also from the point of view of the broader issue of labour absorption in agriculture. Given the rates of growth of the working force and of employment in the nonfarm sector, a structural transformation of the economy would be unattainable in a forseeable future. The number of workers in the rural sector would continue to grow in absolute terms for decades to come, and these workers have to be absorbed in agriculture itself. Now that the seed-fertilizer revolution has made it possible to raise output, income and employment on small farms considerably over past levels, in these analysts' view, such a strategy has become eminently feasible.

While in the initial years of the green revolution such an evaluation of its employment and income potentials looked plausible, it no longer appears to be so in the light of the experience acquired over the last few years. It is known now that the scope and the potential of the green revolution are severely limited under existing conditions of Indian agriculture, so that the strategy of unimodal farm size distribution has lost much of its appeal. It is, however, pertinent to question the basic assumption implicit in the strategy that a unimodal distribution of farm sizes would ensure that the operational or management units would remain small or that the current and growing labour force can be absorbed in agriculture.

Figures 6.1 and 6.2, presented earlier, depict the all-India distributions of landowning households and operational farms by size in 1960-61. Both distributions are highly skewed but both are unimodal. And yet, there were some three million operational units ranging from 20 to 70 acres in size and some 8.5 million landless agricultural labour households in 1960-61. Obviously, a unimodal farm size distribution is no guarantee that operational units would remain small or that the current labour force can be absorbed in farming. Additional measures, such as the imposition of an extremely low level of ceiling on landholdings and land redistribution may seem to be indicated, but whether these measures would succeed in keeping all operational units small and in absorbing additional labour in farming is highly uncertain. After all, there is nothing to prevent an operator from leasing-in additional land, unless leasing-in is entirely forbidden.

This uncertainty notwithstanding, let us proceed to examine whether the strategy of labour absorption in agriculture through land redistribution is feasible. The belief that available land is sufficient for this purpose is quite widespread and not necessarily confined to the popular press alone. One commentator for instance has argued that "realization of economies of scale and of specialization in commercial farming sector and the improvement of the population-supporting capacity of the total rural economy are not necessarily mutually exclusive alternatives" in the Indian context.[24]

Whether the alternatives are mutually exclusive or not will of course depend upon how much land is available and how the term population-supporting capacity of the rural economy is interpreted. The availability of land needs to be examined in the context of the states or regions. Otherwise, the results may be wholly misleading. For India as a whole, there were approximately six acres of land on an average per rural household in 1960-61; but to accept this figure as representing the true availability of land would be to succumb to the fallacy of the average. Furthermore, in a strict sense, neither land nor households are transferrable from one place to another; the question of availability of land should, therefore, be considered at least at the state level.

Labour absorption on land, properly interpreted, should mean provision of a "minimum holding" for each rural household—a holding that is large enough to provide full employment to family labour and to a pair of bullocks, and a minimum income sufficient for family subsistence. Otherwise, labour absorption would degenerate into make-believe self-employment, or make-believe absorption. Several studies have indicated that under Indian conditions a five acre (unirrigated) land holding

[24] Wyn F. Owen, "Discussion : Implications of the Green Revolution for Economic Growth," *American Journal of Agricultural Economics*, December 1970.

may just about provide a farm family of average size with adequate employment and a subsistencein come.[25] Accordingly, we shall take five acres as the absolute minimum size of holding that needs to be assured to each rural household.

It is true that the green revolution has reduced the threshold of subsistence; but it has done so only for the *fully and perennially irrigated farms*.[26] For unirrigated land, the threshold of subsistence still remains five acres. Since the proportion of irrigated area is small in India, and since the bulk of the irrigated land is only seasonally irrigated and dependent upon uncertain and variable monsoon for water supply, we shall treat all land as unirrigated and take five acres as the minimum size for our purposes here.

There is yet another reason for accepting five acres as the minimum holding size. Although the minimum size for subsistence would vary somewhat from place to place depending upon land quality, availability of irrigation and the like, from the point of view of the programme seeking to absorb labour on land this variation is not relevant. If past experience is any guide, the surplus land generated by a ceiling legislation would be of extremely poor quality, and perhaps an area greater than five acres would be needed for allocation to each of the landless families so that it is possible to subsist on land. At the same time, it would be difficult for a programme of this kind to deny the minimum holding to the large number of mini and small farmers whose land base is not adequate to provide sufficient work and/or income. No matter which way one looks at the question, the

[25] V. M. Jakhade and N. A. Mujumdar "Subsistence Sector in Indian Agriculture," *Reserve Bank of India Bulletin*, August 1963; A. M. Khusro, "An Approach to Farm Planning Among Small Farmers," in *Problems of Farm Production Planning and Programming*, Seminar Series 4, Indian Society of Agricultural Economics, Bombay 1964; A. M. Khusro, "Land Ceiling Laws—II," *The Statesman*, New Delhi, June 21, 1972.

[26] Recent studies on small farms initiated in several locations by the All India Rural Credit Review Committee and the Planning Commission suggest that 2.50 acres of irrigated land is the minimum required for viability. These studies have been summarized in B. Venkatappiah, "Small Farmers Development Agency: Outlines of a Programme of Action," address delivered at the 29th Annual Conference of the Indian Society of Agricultural Economics, Waltair, December 30, 1969. Dantwala, however, holds that 'the new technology has reduced the threshold of nonviability to something like 3 irrigated acres," M. L. Dantwala, "From Stagnation to Growth," *op. cit.*

programme must generate enough land so that a minimum holding of five acres can be assured to each household.

A five-acre holding may appear much too large specially in comparison with the average size of the small farms in Japan. It may be asked, if the Japanese small farmer can produce enough from two acres of land to have a high standard of living, why do the Indian farmers need five acres? The answer to this question is partly that the small farmers in Japan and in India are not comparable. Part of the high value productivity of the Japanese small farmer is due to the price of rice which has been arbitrarily kept at a very high level; partly it is due to the fact that over 60 per cent of the arable land in Japan is irrigated and the precipitation is about uniformly distributed over the growing season. Under a similar set of conditions it would perhaps be possible for the Indian farms of identical size to raise productivity to comparable level.

But that is not all. Productivity and income of the average small Japanese farm, though high, is not sufficient for subsistence, much less for a high standard of living. In fact, the small farm operators derive only partial subsistence and partial employment from land. Their farm income need to be supplemented by income from nonfarm work. The stereotype of small but prosperous Japanese farmers has no foundation in reality.

According to Sawamura, "The farm income of Japanese farmers for the most part does not suffice for the household expenditure."[27] Quoting survey data for the Hokuriku prefecture, where the percentage of farmers with side jobs is close to the national average, he points out that income from farming on the average came to about 74.3 per cent of household expenditure in 1954, and this percentage was directly related to farm size; for farms with less than 1.23 acres of land this percentage formed only 29 per cent and it increased to about 98.5 per cent for farms cultivating more than 4.9 acres of land. Viewed from a slightly different angle, the percentage of nonfarm income to total income is the highest in the case of small farms and is inversely related to

[27] Tohei Sawamura, "The Household as a Factor in Farm Planning," paper presented at the Farm Management Development Centre for Asia and the Far East, 14 October-9 November 1957, published in *Papers on Farm Planning and Management*, Government of India, Ministry of Food & Agriculture, New Delhi 1959.

farm size. For Kinki prefecture, Sawamura's data show that for farms with less than 1.23 acres of land, nonfarm income formed 69 percent of total income and 73 per cent of household expenditure.

Sawamura's data clearly underscore the fact that the average small Japanese farm is a part-time, not a whole-time farm. Without an expansion of nonfarm employment, therefore, it seems unreasonable to expect that small farms in India could survive with anything less than five acres of land.

To return to the main issue, three sets of estimates have been worked out from the 1960-61 NSS data on ownership holdings. Column 1 in Table 6.2 shows the estimated size of holding on which ceiling needs to be imposed; since in many states available land is insufficient to provide five acres to each rural household, estimates of additional land required in these states for this purpose are provided in column 2; finally, since additional land is unavailable, estimates of the number of 1961 rural households which must be transferred from agriculture to other sectors of the economy are presented in column 3. There are in all four

TABLE 6.2 : CEILINGS LEVEL AND ADDITIONAL LAND REQUIRED
TO PROVIDE MINIMUM OF FIVE ACRES TO ALL
RURAL HOUSEHOLDS

States	Required ceilings level (acres)	Additional land required (million acres)	Number of households to be shifted out (million)
	1	2	3
Andhra Pradesh		5.110	1.02
Assam		5.593	1.11
Bihar		18.880	3.77
Gujarat	12.50		
Kerala		9.332	1.86
Madhya Pradesh	15.00		
Mysore	10.00		
Orissa		6.757	1.35
Punjab		0.390	0.07
Rajasthan	50.00		
Tamil Nadu		21.729	4.34
Uttar Pradesh		20.900	4.18
West Bengal		13.192	2.63

states where cultivated area is adequate to provide each rural household a minimum of five acres of land; these are Gujarat, Madhya Pradesh, Mysore and Rajasthan. But in nine out of thirteen states available land is not sufficient for the purpose.

In Andhra Pradesh, for instance, a total of 33.20 million acres of land would be needed to provide five acres to those households which were either landless or which did not possess five acres. Even if a ceiling were imposed on five acres of land, the required quantum of land could not be generated as surplus. An additional area of 5.11 million acres must be found so that all rural households could get five acres of land. We can also interpret the estimate in the following way : unless some rural households are transferred to urban areas, or are absorbed in nonfarm occupations or are excluded from the purview of the land redistribution programme, a five acre minimum size of holding cannot be provided to each rural household in Andhra Pradesh; the number of rural households so excluded would be about 1.02 million. In the case of Kerala, if available land were reallocated equally to all rural households, each would receive only 1.2 acres of land; to provide five acres of land to each household, 9.32 million acres of additional land will have to be found, or alternatively, 1.86 million rural households (or about 75 per cent of rural households) must be excluded from the purview of the redistribution programme. A similar interpretation holds for seven other states. In all, about 102 million acres of additional land must be found in nine states, or else about 20 million rural households must be rehabilitated outside agriculture. It is hardly necessary to add that to accommodate the number of new households since 1961, the requirement of additional land would be considerably higher; alternatively, a greater number of households must be rehabilitated elsewhere.[28]

The conclusion is, therefore, inescapable that there is not sufficient land to absorb gainfully the entire landless work force in farming. Certain conditions for labour absorption in agriculture emerge very forcefully from this discussion and they

---

[28] If the entire cultivated land were irrigated, even then about 10 million acres of additional land would be needed in four states (Assam, Kerala, Tamil Nadu and West Bengal) to assure each rural household of a minimum holding of 2.50 acres. Alternatively, about four million households in these states would require to be rehabilitated outside agriculture.

may be spelled out here. One of these is that the area under irrigation be expanded at a rapid pace and to the fullest possible limit; it is equally important to improve the quality of the existing irrigation network. The Indian experience suggests that labour use may be doubled per acre and output per acre may increase by a factor of 1.7, if irrigation is available, even without the high yielding varieties; as water resources are developed, additional farms and additional labour can be productively accommodated in agriculture. Without a rapid expansion of irrigation facilities, therefore, it is futile to talk about absorbing additional labour on unirrigated land with uncertain and inadequate rainfall.

Development of water resources is a necessary but not a sufficient condition for labour absorption however. As the Japanese illustration underscores, the second major condition for labour retention in agriculture is that there must be a vigorous and sustained growth of nonfarm occupations to absorb fully the new entrants to the labour force and at least partially the farmers with palpably small plots of land. Absorption of additional labour in Japan took place not in the farm sector, but in the nonfarm sector. As Ohkawa and Rosovsky observed, the young workers left the farms singly for work in the nonfarm sector and the expansion of nonfarm employment was at a rate sufficient to absorb the new entrants to the work force.[29] In fact, the growth of the nonfarm sector was additionally large enough to provide the farmers with productive and remunerative employment on a part-time basis. It is for this reason that a large proportion of workers could be retained on farms with full employment and a comfortable income.

If these two conditions remain unfulfilled, the unimodal farm-size strategy would at best expand make-believe self-employment in agriculture. Poverty and unemployment would continue to grow, however, since the new entrants to the work force would have no place to turn to for productive employment and since the farmers would have no means to supplement their meagre income from the land.

---

[29]Kazushu Ohkawa and Henry Rosovsky, "The Role of Agriculture in Modern Japanese Economic Development," *Economic Development and Cultural Change*, Vol. 9, Part 2, October 1960.

Chapter Seven

BEYOND THE GREEN REVOLUTION
A Summing Up

The high yielding varieties appeared on the Asian scene at a time
when the Malthusian spectre of famine and hunger seemed to
loom large over the horizon and it appeared that the less developed countries were fast losing the capacity to feed themselves.[1]
Almost at once, the mood of despair seemed to give way to
one of exuberant optimism about the food-population problem.
The high yielding varieties were widely acclaimed as Cornucopia
that would ensure an abundant supply of food. The most favourable construction came to be placed on some of the attributes
of the high yielding varieties, such as the attributes of divisibility,
scale-neutrality and labour intensity; because of these properties,
the green revolution appeared to hold the key to the solution
of the basic problems of Asian agriculture—poverty and unemployment.[2] These expectations were based on a crucial assumption that the scope of the green revolution would be unlimited.

In the light of the experience of the last eight years, these
expectations, for the most part, appear to have been unwarranted.
The error lay in overlooking the implications of the technical
properties of the high yielding varieties and the diverse milieu
in which the new varieties were to be introduced. This led to a
gross miscalculation of the scope of the green revolution itself.
Given the dependence of the high yielding varieties on controlled

[1]William and Paul Paddock, *Famine 1975!* Weidenfeld and Nicholson,
London, 1968.

[2]These expectations at least had some support in the attributes of the
high yielding varieties. Others defied common sense. One example is the
expectation that the spread of the seed-fertilizer revolution would crown
the efforts to spread family planning with a rapid success. See, John Cownie
Bruce F. Johnston and Bart Duff, "The Quantitative Impact of the Seed-
Fertilizer Revolution," *op. cit.*

application of water, and also given the limited development of water resources, the area that can be planted with these varieties is severely limited. Additionally, unsuitability of some of the varieties to specific agronomic conditions in different locations further restricts the scope of the green revolution.

It is, therefore, not surprising that in India the anticipated abundance of food supply failed to materialize. True, despite the limitations, aggregate foodgrains output increased considerably between 1967-68 and 1970-71, and with a foodgrains stock of about 9 million tonnes held by the public agencies in 1971, self-sufficiency status in respect of domestic requirement of foodgrains appeared to have been achieved by India. But one should not exaggerate the significance of the size of the accumulated stock in 1971, since the accumulation was made possible at least partially by the import of about 15 million tonnes of foodgrains between 1968 and 1971.[3]

The scope of the green revolution being thus limited, the expectations concerning reduction in unemployment and poverty were also unwarranted. There have been some areas where, irrigation being satisfactory, demand for labour, employment and wages have risen and others where, owing to lack of irrigation, such increases are not perceptible at all. This is of course inevitable, given the distributions of the limited area under irrigation and of the labour force.

The claims for the green revolution were, therefore, highly exaggerated. But there is no denying the fact that the high yielding varieties have brought about a change in Indian agriculture. It is to the credit of these varieties that they improved employment and income at least in some areas, and that they made a perceptible impact on aggregate food output. Without the high yielding varieties, the problem of poverty and unemployment would have been more acute; without them, there may have been a deceleration in the growth rate of the output of some foodgrains.

On the other hand, predictions of an emerging pattern of adoption biased toward the affluent operators of large farms, premature tractorization displacing labour and tenants, culminating in agrarian tensions and violent conflicts were equally

[3]In 1971, the foodgrains imports totalled slightly more than 2 million tonnes.

unwarranted. If the green revolution has not turned out to be a Cornucopia, neither has it been a Pandora's Box. Adoption of the new varieties has not been confined to a few affluent operators; instead, a broad spectrum of farms in which the small and the medium sized farms predominate, are using the new varieties. Widespread adoption of this nature is due not so much to the scale-neutrality of the high yielding varieties, but partly to the inherited pattern of distribution of irrigated land and partly to the government initiated mass action programme under which all irrigated land and all irrigated farms—small, medium and large—are sought to be covered. Although in some regions and for some crops the operators of large farms may have led in the adoption of the new varieties, small and the medium sized farms have quickly followed; in other areas and for major grains it is the operators of the small and medium sized farms who have been the first to adopt the new varieties. Judging by the distribution of irrigated land, the benefits are biased neither against the small farms, nor in favour of the large farms, but in favour of the medium sized farms. Tractorization of farm operations has been confined by and large to a limited area in few states. There is as yet no evidence that this limited tractorization has led to a large-scale eviction of tenants or to a displacement of hired labour.

It is in this setting that the fancied effects of the green revolution on the rural society, the prospects of polarization and violent agrarian conflicts need to be viewed. As for interregional disparities, these may be striking, but there is nothing to suggest that these disparities are of a destabilizing nature or that these could tear the country apart.[4] Regional growth of agriculture had been disparate even before the green revolution appeared on the scene.[5] The differences between Punjab and Assam, for example, or between Tamil Nadu and Uttar Pradesh in respect of crop output growth rates in the period 1952-53 to 1964-65 were perhaps no less striking than they are now;[6] what is more,

[4]See, Walter Falcon, *op. cit.* The example of Pakistan in this context appears to be particularly inappropriate.

[5]See, W. E. Hendrix and R. Giri, *India's Agricultural Progress in the 1950's and 1960's*, Government of India, New Delhi, 1970.

[6]Compound annual rates of growth of all crop output in Punjab, Assam, Tamil Nadu and Uttar Pradesh during this period were 4.56, 1.17, 4.17 and 1.66 respectively.

differences between districts located within the same state, such as between Bhatinda and Gurdaspur in Punjab, were often amazingly large.[7] But this disparate growth was hardly a cause of tension either among states or among districts. It is indeed questionable if the prosperity of a group of farmers in one part of the country is necessarily a matter of frustration and envy to farmers in another part.

Likewise, the consequences of intraregional disparities too seem to have been magnified out of all proportion. Even the much-publicized Thanjavur incident of 1968 is poor evidence that the green revolution was about to turn red;[8] and the connection that was sought to be established between the green revolution and agrarian conflicts now appears to be questionable in the light of past experience. Scrutiny of the reported cases of conflicts and incidents of forcible occupation of land would show that the relationship between the two is tenuous and more fancied than real. The states with the largest number of recorded disputes happen to be those where agriculture has remained virtually stagnant, or where the progress of the green revolution has been tardy on account of technical problems, such as in predominantly rice-growing areas.[9] But such conflicts have been few in the wheat-belt where the green revolution has had the greatest impact. While tensions and conflicts do exist in the Indian villages, as they always did, there is no evidence of any significant increase in polarization and agrarian unrest that could be attributed to the green revolution.[10]

[7]W. E. Hendrix and R. Giri, *Regional Differences in Crop Output Growth in Punjab* : 1952-53—1964-65, Government of India, New Delhi, 1967. Compound rates of growth of all crop output in Bhatinda and Gurdaspur were 7.85 and 1.24 respectively during this period.

[8]For an analysis of the agrarian relations in Thanjavur, see, Andre Betaille, "Agrarian Relations in Tanjore (Thanjavur) District, South India," paper presented at the Conference on Tradition in Indian Politics and Society, School of Oriental and African Studies, London, July 1-3, 1969.

[9]T. K. Oommen, "Green Revolution and Agrarian Conflict," *E & PW*, June 26, 1971.

[10]Peter von Blanckenburg, "Progressive Farmers and Their Share in Agricultural Modernization : Preliminary Report on a Study in Mysore and Punjab" (mimeographed), Bangalore, July 1972. A summary of the study has been published in *E & PW* (September 30, 1972) as "Who Leads Agricultural Modernization? A Study of Some Progressive Farmers in Mysore

In an important sense, the problems of equity and welfare are enduring problems of Indian agriculture. Redistributive measures did not succeed in resolving them in the past and are unlikely to do so in the future. The reason these problems have remained largely intractable is that the economy has not been growing fast enough to provide the growing number of workers with employment and increases in real income. As Lewis observes, the chances of tackling these problems effectively are much better when the economy is growing at a rate of, say, three per cent per capita per annum rather than at a rate of one per cent or less per capita per annum.[11] Aggressive measures are now needed to promote growth both in the farm and the nonfarm sectors.

The real significance of the high yielding varieties lies in the fact that these provide a base for further development of agriculture. Agriculture's growth in the past was limited by the absence of crop varieties that could respond to modern inputs like fertilizer in terms of grain yield. Now, such varieties are available for some crops; to make their use extensive, measures need to be adopted to ensure an adequate supply of modern inputs like fertilizer and pesticides, and a rapid development of water resources and of drainage facilities. Further development of agriculture will not be cost-free; massive investment in irrigation and in the production of complementary inputs will be necessary to exploit the opportunities of growth held out by the high yielding varieties.

Some writers suggest that the underdeveloped countries have reaped the benefits of the new technology at no extra cost to themselves. These varieties were developed by scientists through research sponsored and financed by international agencies and organizations like the Rockefeller and the Ford Foundations; all that the underdeveloped countries needed to do was to use the results of this research. This is, however, a partial truth at best; for it ignores the large investment made over the last hundred years or so by the underdeveloped countries for the development of water resources—investment that prepared the

---

and Punjab." See also Gilbert Etienne, "India's New Agriculture: A Survey of Evidence," *South Asian Review*, April 1973.

[11] J.P. Lewis, Statement on the Political, Social, Cultural and Economic Impact of the Green Revolution, Symposium on Science and Foreign Policy, *The Green Revolution*, Washington D.C., 1970.

ground for the successful utilization of the high yielding varieties. Irrigation development in India has a long history. In the north-western region and in Punjab (including the part now in Pakistan)—that is in the area where the new wheat varieties have had the greatest impact—large investments in the development of water resources were made in pre-independence days. (In fact, about 50 per cent of the gross investment in irrigation was in the Punjab alone in the period 1898-1914.) And since 1947, considerable investment has been made in irrigation development and in the expansion of irrigated acreage. It is due to this large investment in the past that the high yielding varieties can be used at all. If this investment is taken into account, the high yielding varieties would not appear to be as cost free a technical change as they have been made out to be.

There is yet another reason for stressing the need for a rapid development of water resources and extension of irrigation. As irrigation expands and farmers shift from dry to irrigated farming, or from monoculture to multiple cropping, a greater number of workers will be able to find employment in agriculture. In fact, extension of irrigated farming represents the most important safety-valve for the growing labour force in agriculture.

The search for disease and pest-resistant new varieties must go on with renewed vigour. Research for developing high yielding varieties suitable for cultivation under rainfed conditions in the *kharif* season has hardly begun; without a breakthrough in this kind of varietal research, it would be difficult to widen the scope of the green revolution. The All-India Coordinated Rice Improvement Project has done a commendable job in developing, testing and extending new varieties; but as yet its efforts have remained confined mostly to irrigated *rabi* rice.

One productivity raising measure that needs to be taken up in earnest is consolidation of fragmented holdings. When a thirteen acre holding is fragmented in eight different parcels with an average area per parcel of 1.66 acres, as the NSS, Seventeenth Round data show, the chances of the farmer's raising yield and income through investment in land development, irrigation and the like are meagre. The problem appears to be worse in the case of smaller holdings; the same NSS data show that on an average a 0.73 acre holding is fragmented into about three plots, each measuring no more than 0.24 acre. Without consolidation of fragmented holdings, investment in agricultural infrastructure is unlikely to be fully productive.

It is unfortunate that the spate of policy prescriptions that has followed in the wake of the green revolution has diverted attention from tasks that should have received the topmost priority—that is, tasks such as an adequate production of critical inputs and the development of water resources and drainage facilities. More attention needs to be given to the task of strengthening of the base of the green revolution. It should be realized that the green revolution is but a frail sapling that needs very careful nurturing. Continued scarcity of needed inputs—fertilizer, pesticide and quality seeds—may adversely affect the area under the high yielding varieties; fields may go dry and yields may decline if power to operate pumpsets is not available on time. Farmers' willingness to grow the new varieties may receive a setback from losses sustained owing to nonavailability of inputs at critical stages; unremunerative prices may easily persuade them to shift land from high yielding varieties to other crops. There is no room for complacency. It cannot be taken for granted that the output of foodgrains will continue to increase in the same manner as it did in the period 1967-68 to 1970-71. The progress achieved thus far can be easily lost if the conditions to sustain it are not fulfilled. In view of the unabated growth of population, there is no option now except to strengthen and to widen the scope of the green revolution.

It is true that the green revolution has not resolved the problems of equity and welfare; but it was unreasonable in the first place to expect it to do so. These problems can only be tackled through an all-round growth of the economy; redistributive measures are unlikely to be of much help. Special programmes to raise output and income on small and mini farms such as the Small Farmers Development Agency (SFDA) and the Marginal Farmers and Agricultural Labour (MFAL) programme could perhaps play a useful role. However, as it turns out, these two programmes do not go far enough to tackle the problems of raising productivity and income on small and mini farms. Essentially, SFDA and MFAL have been conceived as agencies to distribute subsidy and to provide risk funds to commercial banks, land development banks, cooperatives and agro-industries centres. It is true that if properly used, credit could help raise productivity and income on some small farms; but it is naive to believe that provision of ample credit is all that is needed for this purpose. The amount of credit that a small farmer can profitably use is small and limited, and very few farmers, cr for that matter, landless labourers can make use of credit to set up profitable enterprises like poultry-farming and dairying. To raise productivity on small and mini farms, the SFDA needs to be transformed first from a subsidy distributing agency to an integrated farm planning agency. Research has shown that scientific management and farm planning can raise yield and income on small farms and significantly reduce the gap between production and consumption on these farms.[12] Technical and management assistance combined with judicious use of credit may be more important than mere supply of credit.

In an important sense, the success of special programmes depends upon the completion of certain unspectacular tasks. For instance, records of tenancy and of rights in land must exist before credit can be provided for investment. Tenants and sharecroppers are eligible for loans if records are available to establish them as tenants or sharecroppers with regard to the land they cultivate. Such records, however, are either not available, or where they exist, are seldom up to date. A similar problem exists also for the small owner-operators. Village-wise records do not often enlighten one as to the size of an individual's total holding that may be scattered over more than one village. Finally, the extent to which small farmers would be able to avail themselves of long-term credit for land improvement and irrigation would depend upon whether or not fragmented holdings have been consolidated. Development and improvement of land records and consolidation of fragmented holdings may be time-consuming, unspectacular tasks, but their completion appears to be a precondition for the success of special programmes like SFDA.

The problems with regard to the MFAL programme are, additionally organizational. The programme has to ensure supply of high quality cross-bred milch animals, feed and fodder

[12] Bains drew attention to the results of research carried out at the IARI on scientific farm management. See, S. S. Bains, "Improving the Production Potential of Small Holdings," in *Problems of Small Farmers*, Seminar Series 7, Indian Society of Agricultural Economics, Bombay 1968.

on the one hand, and on the other resolve the problem of marketing of milk and milk products. Organization is a scarce input anywhere, and specially so in India. But once these problems are resolved, the programme seems to hold a promise for bulk cf the poor in the rural areas.

These production oriented programmes should be distinguished from others which are conceived primarily as relief or welfare programmes, such as, the rural works programme. These are not generally productive, and are unlikely to add to the productive capacity in the rural areas commensurate with the outlay. In view cf the limited resources, the size of such a programme must necessarily be small; consequently, even as a relief measure, such a programme is unlikely to make any significant impact on rural unemployment and underemployment.

The cost of an ambitious programme to provide whole-time year-round jobs on a continuing basis to the unemployed and the underemployed rural workers has been recently worked out by Dandekar and Rath.[13] According to their estimate a recurrent annual outlay of about Rs. 1,028 crores would be needed for the purpose. By any standard, this is a massive outlay representing about twice the total annual public outlay on all agricultural programmes, or about 1.6 times the annual outlay on all industrial and mineral programmes in the central sector provided for in the Fourth Plan document. Whether the country can afford to undertake a programme of this magnitude on a continuing basis is a question that need not hold us here. But recurring outlay of this magnitude on a rural works programme is not wholly a matter of financial programming. There must exist a large amount of surplus, unutilized resources in the economy—resources such as iron and steel, cement, bricks, other road building and construction materials, tools and equipment. Even if the use of physical resources could be kept to the minimum, the size of the programme would necessitate the transfer of large quantities of these resources from other uses. This diversion of resources from use in other sectors may lead to a drastic lowering of the already sluggish rate of growth of the entire economy. Besides, there is also the question whether the opportunity cost of investment in such a programme would not be very high and whether

[13] V. M. Dandekar and Nilkant Rath, "Poverty in India," *E &PW*, January 9, 1971.

the same investment in other sectors or in other projects would not have a larger pay-off at least in terms of immediately consumable output. Although the quantum of employment generated by other projects may be less, such employment would be on a long-term basis and would at least be self-sustaining.

Furthermore, without real resources growing concurrently, the effect of this kind of programme is likely to be highly inflationary. The projects that could be included in such a programme would add little to the total product immediately; their effects on output, if any, may be felt after a long lapse of time. Meanwhile, the rest of the economy must be able to generate the real resources on which the newly-created income would be spent. The assumption that the economy can raise the output of real resources in the required manner seems to be questionable in the light of the record of the economy in the last two decades. So far as foodgrains are concerned, an impressive increase in output has taken place only in wheat; growth in output of jowar and bajra, the items on which a large part of the newly created income is likely to be spent, has not been large. Output of pulses has remained stagnant while that of edible oils has lagged far behind the growth in demand. Other consumer goods too are in short supply and their rates of growth are too small to sustain a massive expenditure of this magnitude.

To undertake a programme of this kind, real resources must increase very substantially. But increase in real resources would require increase in production goods as well. In other words, there must be an all-round growth of the economy. But then, if there is growth of this magnitude, a massive public works programme may be superfluous, for in the process of growth, the bulk of the unemployed and the underemployed would be drawn into productive employment.

Thus, there seem to be no short-cuts, no easy solutions to the problems of rural poverty and unemployment; nor can these be effectively tackled by measures initiated within the confines of the farm economy alone. For this purpose, employment in the nonfarm sector must expand. If the growth of employment opportunities in the nonfarm sector is adequate enough to at least absorb the new entrants to the work force, the problem of rural poverty and unemployment can be contained or stabilized; if, in addition, it is vigorous enough to absorb a part of the mini farmers and landless workers in part-time employment, the magnitude of the problem can be progressively reduced.

The problem in India has been that despite two decades of planning, expansion of nonfarm employment has been tardy. One index of economic growth is the decline in the proportion of labour employed in the agricultural sector. At the time of the formulation of India's Second Five Year Plan, knowledgeable observers were confident that there would be an absolute decline in agricultural work force by 1966, and that a 13 per cent decline in the proportion of agricultural work force would be feasible by 1971.[14] Expectations at the time of the formulation of the Third Plan were that by 1975-76, the proportion of agricultural labour force to the total should come down to 60 per cent or so.[15] These expectations have been belied by subsequent development. The proportion of working force in agriculture declined from 70 per cent in 1951 to only 69.8 per cent in 1971.

A consensus seems to be growing that the rates of growth of employment and output in the nonfarm sector cannot be faster than the rates registered in the past; therefore, it is the farm sector that must somehow absorb additional workers. But it does not appear to be legitimate to assume that the growth of the nonfarm sector cannot be improved upon or that nonfarm employment cannot be expanded enough to absorb rural workers. A corollary of economic development is supposed to be a shift in the locus of activities from the farm to the nonfarm sector and in this process nonfarm employment is supposed to absorb an increasing number of rural workers. It is true that an absolute decline in the number of farm workers takes time to set in, but it is also equally true that at least the relative proportion of workers in agriculture should start declining from the early phases of development. If this decline does not take place and if employment and output in the nonfarm sector do not grow at a faster pace, then regardless of what measures are taken in the farm sector, the problems of poverty and unemployment, of equity and welfare will remain intractable.

[14] See, V. K. R. V. Rao, "The Second Five Year Plan : Employment Pattern and Policies," in Government of India, Planning Commission, *Papers Relating to the Formulation of Second Five Year Plan.*

[15] Government of India, Planning Commission, *Third Five Year Plan,* p. 157.

AUTHOR INDEX

SUBJECT INDEX